A Problem Based Guide to Basic Genetics

to accompany

BIOLOGY

fifth edition

SOLOMON BERG MARTIN

Donald Cronkite

Hope College

Saunders College Publishing

Harcourt Brace College Publishers

Fort Worth Philadelphia San Diego New York Orlando Austin
San Antonio Toronto Montreal London Sydney Tokyo

Printed in the United States of America

ISBN 0-03-022109-9

901 202 765432

CONTENTS

Solving Genetics Problems

Solving genetics problems is not very difficult if you are systematic in your approach. The trick is to recognize the inheritance pattern that each problem contains. Only a few basic patterns exist, so once you have learned to spot those patterns, you will find that you can recognize most any problem you are presented. That's why you need to do many genetics problems to learn genetics. The more problems you do, the more practice you will get at recognizing the patterns.

Chapter 1
Seven Systematic Steps for Solving Genetics Problems

1. Always use standard designations for the generations and lay out the problem as a simple diagram.

When you cross two organisms, they are called the parents or the P generation. Their offspring are called the first filial generation or F_1. If you cross two F_1 individuals, their offspring are called the second filial generation or F_2. These terms are always used in genetics problems; understand the terms and use them whenever working a problem.

Many problems are stated in paragraph form, often covering up the basic pattern with words. You will find it easier to do most problems if you change the information into a simple diagram of the crosses, making the patterns much easier to recognize.

Solved Problem 1-1. Although flies usually have a pair of wings, geneticists have found some wingless strains. When one of these wingless flies was crossed to a normal fly with wings, all of their offspring had wings. When some of these offspring were crossed to each other, they produced 428 offspring, of which 320 had normal wings and 108 were wingless.

Summarize this cross in a diagram, using the correct terms for each generation.

Answer: The three generations can be summarized in this way:

P: winged X wingless

F_1: all winged

F_2: 320 winged, 108 wingless

Practice Problems (All answers to practice problems begin on p. 64.)

1-2. A rare white bison was born in a bison rancher's herd. When it matured, the rancher mated it to a normal bison on three occasions, and each time a normally colored bison was born. When those new bison grew up, two of them were mated to each other. All of their calves were brown except one, which was albino like its grandmother. Summarize this cross in a diagram, using the correct terms for each generation.

1-3. Silk worm blood (or "hemolymph") is either deep yellow or white. The caterpillar may be plain or heavily marked with stripes and spots. A moth with yellow hemolymph that had been a plain caterpillar was crossed to a moth with white hemolymph that had been a marked caterpillar. Among the offspring were 65 yellow, marked; 56 yellow, plain; 61 white, marked; and 59 white, plain. Diagram this cross using the correct terms for each generation.

1.4. In a certain locality almost all lacewings (an insect of the order Neuroptera) had golden eyes, but a small number had green eyes. When a golden-eyed lacewing was crossed to a green-eyed lacewing, some golden-eyed and some green-eyed lacewings were found among the offspring. Diagram this cross.

2. Write down a key for the symbols you are using for the allelic variants of each locus.

In order for genetics crosses to be done, gene differences have to be identified. Little would be learned by crossing red-flowered plants generation after generation except that there were no gene differences. An important insight of Mendel's was that the genes act like particles that are passed on from generation to generation, and those particles may occur in more than one alternative form. We summarize this idea by designating each allele (alternative form of a gene) with a symbol and all the alternative forms of the same gene (a locus) with related symbols.

The simplest symbol system designates dominant alleles with an upper case letter and recessive alleles with the same letter in lower case. (Specialized notation is used by scientists studying particular organisms, such as *Drosophila*. Simple notation is sufficient for this book, however.) For example, in humans the recessive allele for sickle cell anemia is symbolized by *s*, and the allele for the normal allele is *S*. The symbols only tell us which allele is dominant and which recessive, not which one is "normal" or the most frequent. For example, the dominant mutation "Bar eye" in fruit flies reduces the number of facets in the compound eye. Flies with normally shaped eyes have the recessive allele of this locus, so Bar eye is *B* and normal eye is *b*.

When working with diploid organisms, each individual will contain two alleles for each locus. If the individual is homozygous, we symbolize it with two of the same allele symbol, but if heterozygous, we use two different allele symbols. Gametes are designated with single allele symbols for each locus. If more than one locus is being symbolized in one individual, separate the symbols of different loci with a space.

Solved Problem 2-1. Mendel studied seven different loci in peas, with two alleles possible at each locus. Let tall be *T* and short be *t* and red flower be *R* and white be *r*. Then write the symbols for each of the following.

A) a homozygous tall plant
B) a plant heterozygous for flower color
C) a short plant with white flowers
D) a plant heterozygous at both loci
E) a gamete with the tall allele
F) a gamete with the short and red alleles

Answer. A) *TT* B) *Rr* C) *tt rr* D) *Tt Rr* E) *T* F) *t R*

Solved Problem 2-2. *Drosophila* usually have round eyes, but a recessive mutation results in eyeless flies. They usually have lovely straight bristles on their backs, but a recessive mutation results in crooked bristles that are twisted like little corkscrews. Finally, a dominant mutation changes the normal brick red eyes to a purple color called "plum." Symbolize a fly heterozygous at all three loci.

Answer. You will need to decide what letter you want to use for each locus. Suppose you use these symbols:

E = round; *e* = eyeless
C = straight; *c* = crooked
P = plum; *p* = red

Then the fly in question would be *Ee Cc Pp*.

PRACTICE PROBLEMS

2-3. Huntington's disease is a human malady due to a dominant allele that is not expressed until a person reaches middle adulthood. Cystic fibrosis is a malady due to a recessive allele of a different locus. Choose the letters you want to use and then write the symbols for each of the following individuals.

A) A young person who is destined to have Huntington's disease later in life.

B) A person who does not have alleles for Huntington's disease or cystic fibrosis.

C) A person who has cystic fibrosis, but will not develop Huntington's disease.

2-4. A homozygous fly with an ebony body color (recessive allele) was crossed to a fly homozygous for the wild-type straw-colored body (dominant), and all their offspring were heterozygous for this body color locus. Write symbols for each parent and for their offspring.

2-5. The golden-eyed lace wings in the P generation of Problem 1-4 are heterozygous at a locus for eye color, and the green-eyed parents are homozygous for the recessive allele. Choose a letter and write symbols for the golden-eyed and green-eyed parents.

3. Determine the genotypes of the parents of each cross.

Geneticists deal with two general features of organisms. The genotype is the collection of genes that an organism has, while the phenotype is a description of the characteristics of an organism. Most simple genetics problems are of two general kinds. One, covered in this section, requires you to find the genotypes or phenotypes of parents, while the other, covered in sections 4-7, requires you to find the genotypes or phenotypes of offspring. Three kinds of evidence can be used to figure out the parents.

A. Are they from true-breeding lines?

If a line results in the same kind of offspring generation after generation, they are true breeding. **Such a line must be homozygous.** Sometimes a problem will simply say that the lines used are true breeding. Sometimes other data allow you to figure that out.

Solved Problem 3-1. Which of the following individuals are probably homozygous?

A. A rat from a true-breeding strain of laboratory animals.

B. Two parents with brown eyes who produce a child with blue eyes.

C. A snapdragon with purple flowers from a strain that had consistently produced purple flowers for 14 generations.

Answer.
A. If the rat is from a true-breeding strain, that means it is, by definition, homozygous.

B. The parents are not true breeding, so at least one of them is not homozygous, and probably neither is.

C. "True breeding" means that the strain produces the same kind of individuals in every generation, so this plant is from a true breeding strain and thus is homozygous.

PRACTICE PROBLEMS

3-2. Jake makes a deal with his father-in-law Lavern. They divided up Lavern's flock so that Jake got all the spotted goats and the black sheep while Lavern got the solid colored goats and the white sheep. They agreed that whenever a spotted goat or black sheep appeared in Lavern's flock, he would give it to Jake, and whenever a solid goat or a white sheep appeared in Jake's flock, he would give it to Lavern. As goats and sheep were born in the flocks, Lavern had to keep giving some of the offspring from his flock to Jake, but Jake didn't have to give any of his flock to Lavern. Explain using the terminology of this section.

3-3. Sometimes pairs of wild foxes produce one or more silver-backed pups. Two such pups raised in captivity always have silver-backed pups when mated to each other. Which individuals in this example are homozygous?

B. Can their genotypes be reliably deduced from their phenotypes?

A dominant allele is one that is expressed whether it is homozygous or heterozygous, while a recessive allele is only expressed when homozygous. If we know which allele is dominant and which recessive, we can know the genotype of recessive parents with certainty. If we know nothing about an individual except that it has the dominant phenotype, then its genotype is ambiguous -- it could be either homozygous or heterozygous, and other information will be needed to know which it is for certain.

Solved Problem 3-4. Look at the cross described in Problem 1-1. Suppose that lack of wings is recessive and is symbolized with a lower case *w* and the allele for having wings is *W*. On the basis of that information alone, write the genotypes of all the individuals in the parental, F_1 and F_2 generations that you know for certain.

Answer: All the wingless flies are unambiguously ww. On the basis of only our knowledge of the parental phenotypes, we are not able to tell with certainty what the genotypes of winged flies might be.

P: winged (?) X wingless (ww)

F_1: all winged (?)

F_2: 320 winged (?), 108 wingless (ww)

Solved Problem 3-5. Smooth fox terriers sometimes have congenital myasthenia gravis, a muscle disease due to a recessive allele that causes a defect in the nerve-muscle synapse when homozygous. Such dogs almost always die by 6-9 weeks of age. Can we ever know for certain the genotype of parent fox terriers by looking at their phenotype?

Answer. No, we cannot. The recessive homozygotes have myasthenia gravis and die as puppies. So they can never be parents. Parents will not have myasthenia gravis, so they could be homozygous or heterozygous.

PRACTICE PROBLEMS

3-6. Refer to problem 3-2. Assuming those coat colors and patterns are due to single gene differences, can we know any of the genotypes with certainty?

3-7. Peroneal muscular atrophy afflicts humans, beginning when the individual is 10 - 20 years old. A person has peroneal muscular atrophy only if at least one of the parents had it. Assuming that a single gene difference is at issue here, can you tell whether the gene is caused by a dominant or recessive allele based on these data?

C. Do the phenotypes of their offspring provide any information?

A diploid organism provides one allele of each locus to its offspring. The offspring receives one allele from each parent. Sometimes this makes it possible to deduce the genotype of a parent by looking at the offspring. An organism that has a recessive phenotype must be homozygous recessive and must have received one recessive allele from each parent. So each parent must have at least one recessive allele to give to the offspring. If a parent has the dominant phenotype and there are offspring with the recessive phenotype, that parent must be heterozygous.

Solved Problem 3-8. A man's wife had an albino child. Albinism, a lack of melanin pigment, is due to a recessive allele of a single gene. "Neither you nor I is albino," said the man. "Therefore, I am not the father of this child." Is the man correct?

Answer. Let's symbolize the recessive albino allele as *a* and the dominant allele that produces melanin as *A*. Let's diagram the cross and write what we know for sure.

P: melanin (??) X melanin (??)

F_1: one albino child (aa) [we know this genotype because the phenotype is the recessive one]

Because the parents are not albino, they could be either *AA* or *Aa*. A way of symbolizing that is to write their genotypes as *A_* , where the blank could be either *A* or *a*.

P: melanin (A_) X melanin (A_)

F_1: one albino child (aa)

The albino child must have received an albino allele from each parent in order to be *aa*, so that allows us to fill in the blanks in the parents genotypes.

P: melanin (Aa) X melanin (Aa)

F_1: one albino child (aa)

So the man could be the father of an albino child even if he himself or none of his relatives is albino.

Solved Problem 3-9. Refer to problems 3-2 and 3-6. If Lavern knew anything about genetics, he might prefer to kill and eat the ewes (female sheep) in his flock that gave birth to black sheep. Why would that be a good strategy?

Answer. White ewes have the dominant phenotype. If they give birth to black sheep (*ww*), they must be *Ww*, since the offspring get a *w* from each parent. If he eliminates the heterozygotes, he won't have to keep giving sheep away to his son-in-law.

PRACTICE PROBLEMS

3-10. What can we know for certain about the genotypes of the winged flies in Problem 3-4?

3-11. Farmer O'Sullivan has a prize boar named Honey which has won many honors at the State Fair. Several other farmers want to mate Honey to their sows so they can have prize pigs too. But Farmer O'Sullivan knows that some of Honey's litter mates had been runts and that the failure to grow properly was due to a recessive gene. What does this tell us about Honey's possible genotypes?

3-12. What cross could Farmer O'Sullivan do to find out for certain what Honey's genotype is?

3-13. *Paramecium tetraurelia* usually will survive if the concentration of potassium in the medium is as high as 30 mM. However, recessive alleles of two different loci make the animals more sensitive, causing them to die when potassium levels reach about 20 mM. The sensitive allele of one locus is called *k1*, and the sensitive allele of the other locus is *k2*. When either allele is homozygous, the protozoa are potassium sensitive. There must be a dominant allele (*K1* and *K2*) at each locus for the protozoa to show normal resistance. Which of the following genotypes exhibit potassium sensitivity and which will be resistant?

a) *k1k1 K2k2* b) *K1K1 K2K2* c) *K1k1 K2k2* d) *k1k1 K2K2* e)*K1K1 k2k2*

4. Indicate the possible kinds of gametes formed by each of the parents.

Sexual reproduction involves the alternation of fertilization, which produces zygotes, and meiosis, which precedes gamete formation. Your thinking about genetic problems should center on that alternation of fertilization -- gamete formation -- fertilization -- gamete formation in laying out the problem. Then by systematically filling in what we know and using that information to deduce what we don't know, we can solve most any genetics problem. In this section we will work out the gamete types.

A. If it is a monohybrid cross, we must apply the Principle of Segregation.

In a monohybrid cross we focus on a single locus. When there are just two possible alleles, gamete formation is simple. Each gamete will receive one allele. If the parent is homozygous, all its gametes will be alike. If the parent is heterozygous, each gamete will receive only one of the two possible alleles. This is what is meant by the **Principle of Segregation**.

Solved Problem 4-1. What are all the types of gametes formed by individuals of each of the following kinds? What proportion of every type of gamete will each individual produce?

A. *QQ* B. *Ww* C. *mm*

Answer. Remember, each gamete will receive only one allele, and if there are two possible alleles, half the gametes will get each type.

A. All the gametes of *QQ* will be *Q*.

B. Half the gametes of *Ww* will be *W* and half will be *w*.

C. All the gametes of *mm* will be *m*.

PRACTICE PROBLEMS

4-2. What are all the kinds of gametes that chickens from a true breeding white-feathered strain can produce?

4-3. Refer to problem 3-8. What are all of the kinds of gametes that each parent in that problem can produce?

4-4. If a geneticist discovered an individual heterozygous for an allele that killed any sperm that had that allele, could she/he maintain a stock of organisms that had this gene?

B. If it is a dihybrid cross, we must apply both the Principle of Segregation and the Principle of Independent Assortment.

If two or more loci are being studied simultaneously (a dihybrid cross), then the problem might seem a little more complicated, but it really isn't. If you have two loci, you can treat each like a monohybrid cross. Even if there are many more loci than two, you can still do this and have a simpler problem to solve. Each gamete will only get one allele from each locus. In other words, all the loci are acting according to the Principle of Segregation.

In every gamete a particular allele of one locus can be combined with either of the alleles of the other locus. That is what is meant by the **Principle of Independent Assortment**.

The important thing is to be sure that you have accounted for all of the possible combinations of alleles when you figure out the gametes. Two accounting methods make this rather easy.

1. The branch method. Write down the genotype of the first locus, and then make a little branch showing the kinds of gametes expected from that locus. Suppose for example, that we were working with an organism heterozygous for two loci -- *Aa Bb*.

Write a little branch diagram for the first locus like this:

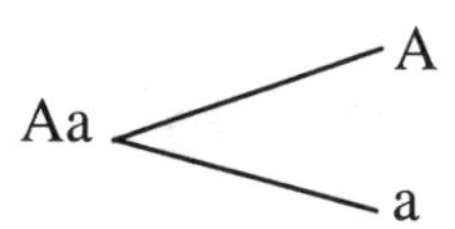

The *A* allele and the *a* allele can now each be combined with either of the choices for the next locus, so write a second little branch diagram next to the end of each of the first branches. Circle the resulting gametes.

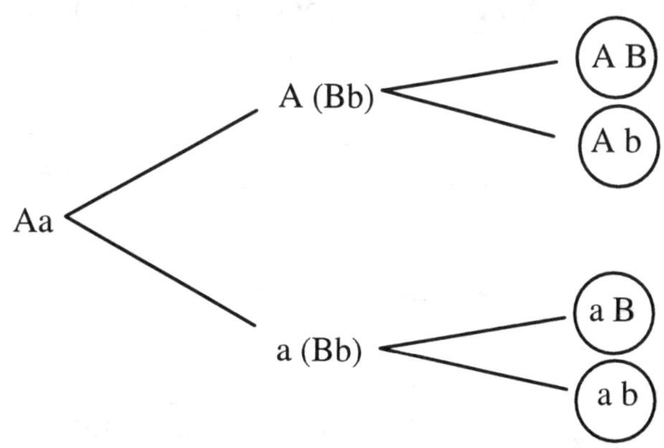

2. The alternating list method. If you know how many gametes you should end up with, then it is easier to be sure you have listed all the possible types. This second method for accounting for all the gametes is based on the observation that the number of different kinds of gametes will be equal to 2^n where n is the number of loci that are heterozygous. Again suppose the organism is the double heterozygote, *Aa Bb*. Because there are two heterozygous loci, there will be $2^2 = 4$ different kinds of gametes. Of the four gametes, two will have an *A* allele and two will have an *a* allele. List these like this:

A
A
a
a

Of the gametes with an *A* allele, half will have *B* and half will have *b*. Likewise, half of those with the *a* allele will have *B* and half will have *b*. So now write a second column on your list like this.

(AB)
(Ab)
(aB)
(ab)

The alleles of the last locus on your list should alternate every other one as in this case if you did it right. Since the list is complete, we've circled them to indicate that they are gametes.

Although we will be concentrating on crosses with only two loci, both the branch and list method will work no matter how many loci you are working with. If you want to be sure you have the idea for the dihybrid cases, try some of the later problems in this section to see if you can do it with a larger number of loci.

Solved Problem 4-5. In each of the following cases use either the branch or the list method to determine the kinds and proportions of gametes that the individual will produce.

A. *Mm Nn* B. *AA bb* C. *RR Tt*

Answer. A. This is exactly like the example except that *M* and *N* have been substituted for *A* and *B*. The four possible types are (MN) (Mn) (mN) and (mn).

B. Remember that the two loci act independently. In this case the first locus is homozygous and can only produce gametes containing an *A* allele. Likewise, since the second locus is homozygous, it too can only produce one kind of gamete, which contains allele *b*. So the only kind of gamete possible is *Ab*. Incidentally, if you use the alternating list method, n = 0 since there are no heterozygous alleles, and the number of different gametes is $2^0 = 1$.

C. To do this one by the branch method, first consider the locus that is *RR*. It will make only one kind of gamete, containing *R*.

RR ———— R

Then the second locus can form either *T* or *t* gametes, so

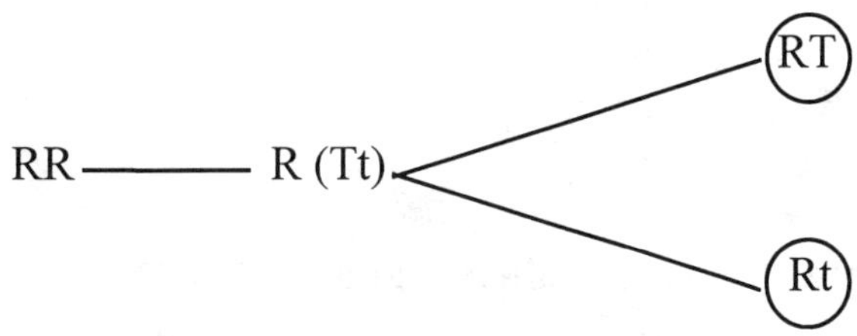

By the list method, first determine the number of possible gamete types, which is 2^1 = 2 since there is only one heterozygous locus. There will be two kinds of gametes, and all will contain R.

R
R

Then half of those will contain T and half t, so

RT

Rt

Solved Problem 4-6. Tay-Sachs disease is a serious human recessive disorder that affects newborn infants who are homozygous for the deleterious allele *t*. People with a single normal allele *T* do not have Tay-Sachs disease. However, they can be detected as heterozygotes for this locus by biochemical screening. A dominant gene *P* produces a characteristic called "polydactyly" in which humans have one or more extra fingers or toes. A person with polydactyly who had one parent with normal fingers and toes is found to be heterozygous for Tay-Sachs disease. What are all the kinds of gametes this person can produce?

Answer. We know the person is heterozygous *Tt* because the problem says so. A person with the dominant trait polydactyly is either *PP* or *Pp*. Because the person has one normal parent (*pp*), that parent would have produced only *p* gametes, so the person under consideration must be *Pp*. The question thus breaks down to "What are all the kinds of gametes a person can produce who is *Pp Tt*?" The answer is *PT*, *Pt*, *pT* and *pt*.

Solved Problem 4-7. Write all the kinds of gametes that can be produced by a person with this genotype: *Aa BB cc Dd Ee ff gg HH.*

Answer. At first this looks rather daunting, but take it one step at a time. Let's use the list method, although either method would work. There are 8 loci in this problem, but only three are heterozygous. So the total number of different gametes that can be produced is $2^3 = 8$. Half will contain allele *A* and half will contain allele *a*. So start the alternating list this way:

A a
A a
A a
A a

Now all those gametes will have a *B* allele and a *c* allele because these two loci are homozygous.

A B c aBc
A B c aBc
A B c aBc
A B c aBc

The next two loci are heterozygous, and in each instance an upper case allele will end up in half the gametes and a lower case allele in the other half. Do the alternating list for each.

A B c D E aBcDE
A B c D e aBcDe
A B c d E aBcdE
A B c d e aBcde

The last three alleles are all homozygous, so the same allele will end up in each gamete. The final list will therefore look like this:

A B c D E f g H a B c D E f g H
A B c D e f g H a B c D e f g H
A B c d E f g H a B c d E f g H
A B c d e f g H a B c d e f g H

PRACTICE PROBLEMS

4-8. Farmer Vandermeer had two hogs named Olga and Hank who were normal healthy hogs, but whenever he crossed them, some of the offspring had cleft palates. Assuming cleft palate has a genetic basis, what are all the kinds of gametes that Olga and Hank can produce?

4-9. If a *Drosophila* is heterozygous for recessive alleles *e* (ebony), *f* (forked), *t* (temperature-sensitive) and *v* (vestigial), what kinds of gametes can that fly produce?

4-10. If an organism were heterozygous for 10 loci, how many different gametes could the organism produce?

5. Set up a Punnett square.

Fertilization follows gamete production, and at fertilization we want to combine the gametes of one parent in all the possible combinations with the gametes of the other parent. A simple way to do that is to construct a Punnett Square. This is a grid with the gametes of one parent written across the top and the gametes of the other written along the left side.

Solved Problem 5-1. Suppose we crossed two organisms each of which had the genotype Bb. Set up a Punnett Square for that cross. What if one parent were *Bb* and the other were *bb*?

Answer. Set out a simple diagram of the cross to show what kinds of gametes each parent will produce.

P: Bb X Bb

Gametes (B) and (b) (B) and (b)

Then construct the Punnett Square.

	B	b
B		
b		

If the parents were Bb and bb, the gametes will be a little different, but the approach will be the same.

P: Bb X bb

Gametes (B) and (b) all (b)

	B	b
b		

Solved Problem 5-2. Set up a Punnett Square for the cross of parents that are both *Aa Bb*.

Answer. Again do a simple diagram of the cross and then use the resulting gametes to set up the Punnett Square.

P: Aa Bb X Aa Bb

Gametes (AB) (Ab) (aB) (ab) (AB) (Ab) (aB) (ab)

	AB	Ab	aB	ab
AB				
Ab				
aB				
ab				

Practice Problems

5-3. Suppose we cross an organism of genotype *MM NN* with one that is *mm nn*. Prepare a Punnett Square for this cross.

5-4. Set up a Punnett Square for a cross of two flies heterozygous for *B*, *F* and *G*.

5-5. If a mouse were heterozygous at 14 different loci, and its mate were heterozygous for just three of those loci, what would be the dimensions of a Punnett square used to study this cross?

6. Fill in the Punnett square and determine the genotypic and phenotypic ratios of the offspring.

With the Punnett Square set up, it is now a simple matter to fill in all the intersections with the alleles from both gametes that form that intersection. In this way you can be sure of having produced all possible gamete combinations. Then, once the genotypes of the new zygotes are determined, you can figure out the phenotypes of each zygote type to complete the problem.

Solved Problem 6-1. Fill in all the intersections on the Punnett Squares of Problems 5-1 and 5-2.

Answer. At each intersection of the square, fill in all of the gametes from the row and column that intersect.

	B	b
B	BB	Bb
b	Bb	bb

	B	b
b	Bb	bb

When Punnett Squares get a little complicated, as in 5-2, follow a systematic pattern for filling in the intersections so you won't get confused.

1) Fill in just one allele at a time, and keep all the alleles of the same locus together. So in the example above, note that all the A's are together in any one intersection and all the B's are together. Fill in the A's first. Then do the B's.

2) When there are two different alleles at an intersection, put the upper case letter first. Notice that there are no genotypes written aA or bB.

	AB	Ab	aB	ab
AB	AABB	AABb	AaBB	AaBb
Ab	AABb	AAbb	AaBb	Aabb
aB	AaBB	AaBb	aaBB	aaBb
ab	AaBb	Aabb	aaBb	aabb

PRACTICE PROBLEMS

6-2. Fill in all the intersections on the Punnett Square of Problem 5-3.

6-3. Fill in all the intersections on the Punnett Square of Problem 5-4.

6-4. Refer to the Punnett Square of Problem 5-5. How many intersections will it have?

7. If you do not need to know the frequencies of all of the expected genotypes and phenotypes, you may use the rules of probability as a shortcut.

Mendel's Law of Independent Assortment tells us that the probability of an allele from one locus going into a gamete is independent of the probability of an allele from another locus going into that gamete. The assorting of alleles into gametes are independent events. There is a rule of probability called the **Product Rule** which says that you can find the probability that two independent events will happen at the same time by multiplying their two individual probabilities. When loci are heterozygous (say *Aa*) , the probability that a gamete will receive the recessive allele *a* is 1/2, and the probability of receiving the dominant allele *A* is also 1/2. So what is the probability of a homozygote, *AA*? It is 1/2 X 1/2 or 1/4.

Then take this one step further. If the parents are heterozygous at two loci (*Aa Bb*), the probability of getting an offspring that is homozygous for both *A* and *B* is 1/4 X 1/4 = 1/16. The “1/4" comes from the reasoning in the previous paragraph.

Solved Problem 7-1. A man is heterozygous for the allele for sickle cell anemia, which is expressed only when homozygous. His wife is also heterozygous for sickle cell anemia. Assuming that male and female children are each born about half the time, what is the probability that this couple will have a son with sickle cell anemia? What is the probability that they will have a normal daughter?

Answer. If both parents are heterozygous for the sickle cell gene, the probability of having a child with sickle cell anemia is 1/4, and the probability of having a child without the disease is 3/4. The probability of having a son is 1/2, so the chance of having a son with sickle cell anemia is 1/4 X 1/2 = 1/8. The chance of having a daughter without sickle cell anemia is 3/4 X 1/2 = 3/8.

Solved Problem 7-2. In cattle red and white coat color is controlled by a recessive allele *r*; the dominant allele *R* causes black and white coat color. Cattle can either lack horns (*HH* or *Hh*) or have them (*hh*). If we cross two double heterozygotes, what will be the proportion of each kind of offspring?

Answer. Consider each locus separately. A cross of *Rr* X *Rr* will produce 1/4 red and white and 3/4 black and white. *Hh* X *Hh* will produce 3/4 hornless (1/4 *HH* and 1/2 *Hh*) and 1/4 horned. To predict the next generation, therefore, we can simply multiply probabilities:

Red and white, horned	1/4 X 1/4 = 1/16
Red and white, hornless	1/4 X 3/4 = 3/16
Black and white, horned	3/4 X 1/4 = 3/16
Black and white, hornless	3/4 X 3/4 = 9/16

PRACTICE PROBLEMS

7-3. Suppose there was a cross in which each parent was heterozygous for 10 different loci. What proportion of the offspring will be homozygous for all 10 recessive alleles?

7-4. Refer to Problem 7-2. A farmer said to a geneticist, "If I cross these 15 heterozygous black and white horned cows to a red and white bull that I'm pretty sure is heterozygous for the hornless gene, and I get two calves from each mating, how many of the calves can I expect to be red and white and hornless like their father?

7-5. Consider Problem 4-7. Suppose two organisms of the genotype in that problem (*Aa BB cc Dd Ee ff gg HH*) were crossed to one another. What proportion of the offspring would have the same genotype as the parents?

CHAPTER 2
GOING BEYOND THE SEVEN STEPS

8. The combination of the seven steps that you use depends on the kind of problem you have to solve.

In general there are two kinds of genetics problems: those that tell you about the parents and ask for conclusions about the offspring and those that tell you about the offspring and ask for conclusions about the parents. These problems require slightly different approaches.

A. Given parental data, make conclusions about the offspring.

This is the most direct kind of problem. Steps 4-7 are most helpful, although you will still benefit from simplifying the problem with a diagram, and you will need to define the gene symbols to be used.

Solved Problem 8-1. The *p* allele of the German cockroach produces a fine mottling all over the body. This condition is called "peppery" and is recessive to the wild type allele, which produces solid coloring. What are the expected genotype and phenotype ratios among the offspring in each of these crosses: A) a cross between two peppery cockroaches. B) a cross between two heterozygous cockroaches. C) a cross between a peppery cockroach and a heterozygote?

Answer. Diagram each cross, put down all that you know, and then construct Punnett Squares to get the answer.

A) P: Peppery X Peppery
pp pp

Gametes all (p) all (p)

Punnett Square:

	p
p	pp peppery

All the offspring are peppery.

B) P: Solid X Solid

	Pp		Pp	
Gametes	1/2 (P)	1/2 (p)	1/2 (P)	1/2 (p)

Punnett Square:

	P	p
P	PP Solid	Pp Solid
p	Pp Solid	pp Peppery

3/4 Solid and 1/4 peppery

C) P: Peppery X Heterozygote (Solid color)

pp Pp

Gametes all (p) 1/2 (P) 1/2 (p)

Punnett Square:

	P	p
p	Pp Solid	pp Peppery

1/2 Solid color and 1/2 Peppery

Solved Problem 8-2. A plant had the genotype *Ff Gg Hh*. Capital letters indicate dominant alleles. If two such plants were crossed, A) what proportion would show the dominant phenotype for F? B) What proportion would show the recessive phenotype for all three genes?

Answer. Since there are 3 heterozygous loci, there will be 8 different possible gametes:

(FGH) (FGh) (FgH) (Fgh)

(fGH) (fGh) (fgH) (fgh)

If you do these problems with a Punnett Square, there will be 8X8 = 64 boxes to fill in, but you don't have to do that given what the problem asks. You can use the product rule of probability.

A) This question is only about *F* and *f*. The loci are independently assorting, so you don't have to consider *G* and *H*. In a cross of *Ff* X *Ff*, 1/4 will be *FF* and 1/2 will be *Ff*. So 3/4 will have the dominant phenotype.

B) In a cross of *Ff* X *Ff*, *ff* will occur 1/4 of the time. Likewise, in a cross of *Gg* X *Gg*, *gg* will occur 1/4 of the time, and in a cross of *Hh* X *Hh*, the *hh* will occur 1/4 of the time. So the genotype *ff gg hh* will occur 1/4 X 1/4 X 1/4 = 1/64 of the time.

PRACTICE PROBLEMS

8-3. Look at problem 8-2. If one of those triple heterozygotes were crossed to a plant homozygous for all three recessive alleles, what would be the proportions of each genotype among the offspring?

8-4. In humans polydactyly is due to a dominant allele and results in extra fingers and/or toes. Phenylketonuria is recessive and is a condition due to a disorder in the metabolism of phenylalanine. Unless given a special diet as infants, people with phenylketonuria may have varying levels of mental retardation. A man who has neither condition but whose father has phenylketonuria and a woman with no phenylketonuria allele but with polydactyly like her father (assume she is heterozygous for polydactyly) wonder what the probability is of having a child with both conditions. What about with one of these conditions?

8-5. Recessive alleles of two different loci in mice give the very same phenotype. One is light ear (*l*) and the other is pale ear (*p*). Each results in ears that are lighter than normal, and each has a dominant allele, *L* or *P*. The loci are independently assorting. Investigators wanted to create a double mutant strain *ll pp*. After a series of crosses, they had a group of light-eared mice, some of which were *ll PP*, some *LL pp* and some *ll pp*. But they all looked alike. They had the original strains of each single mutant as well as the wild type *LL PP*. What crosses can they do to distinguish between the single and double mutants?

B. Given data about progeny, make conclusions about parents.

Here you will work backward. After you have done a number of problems of the first kind, you will see that there are just a few possible ratios that you get among the offspring. Recognizing these ratios is very important because when you know the offspring ratios, that is almost always useful in making conclusions about the parents.

If the offspring ratio is	Then we know this about the parents	The cross is called
3:1	Both parents are heterozygous at one locus. One allele is dominant.	Monohybrid
1:2:1	Both parents are heterozygous at one locus. Neither allele is dominant.	Monohybrid
1:1	One parent homozygous recessive. One parent heterozygous.	Test Cross
9:3:3:1	Each parent is heterozygous at two loci. A dominant allele at each locus	Dihybrid
1:1:1:1	One parent a double heterozygote. One parent homozygous recessive for both loci -OR- One parent heterozygous at the first locus and homozygous at the second One parent heterozygous at second locus and recessive at first i. e. *Bb tt* X *bb Tt*	Test Cross

Solved Problem 8-6. A tropical fish fancier produced an albino zebrafish. It lacked the blue stripes characteristic of the normal zebrafish, and it had pink eyes. When he crossed a normal zebra fish to the albino, all the offspring were normal. When he crossed two of the normal F_1 offspring, 1/4 were albino, and 3/4 were normal. When an albino was crossed to the normal F_1, half the progeny were albino, and half were normal. What is the genotype of each fish mentioned in this problem?

Answer. Look for a cross that gives readily familiar results. There is a three to one ratio among the progeny of the F_1. This is what we expect when two individuals are crossed that are both heterozygous at a single locus. So that cross would be:

Normal (*Aa*) X Normal (*Aa*)

When albino is crossed to normal, the offspring are normal, suggesting that albino is the recessive. So the original parental fish were a normal *AA* and an albino *aa*.

The last cross of albino to the F_1 normal confirms these previous identifications. The 1:1 ratio in that cross is characteristic of a test cross in which one fish (the normal) is *Aa* and the other fish (the albino) is *aa*.

Solved Problem 8-7. A mutation in chickens resulted in extra bones in the wings and feet, a shortened beak and an inability to hatch without help because they were unable to peck their way out of the shell. The mutation is a recessive allele called *d* for "diplopodia." A cross was done involving diplopodia and another locus that controls feather color called "dominant white." The *I* allele results in white feathers, but *i* results in colored feathers.

In a particular cross the offspring included 180 white chickens, 68 with colored feathers and diplopodia, 166 with colored feathers and 53 with white feathers and diplopodia. Figure out the genotypes and phenotypes of the parents.

Answer. Since we know the dominance relationships of these genes, we can write down quite a bit that is known about the offspring without considering the ratios at all:

180 white (no diplopodia)
68 color and diplopodia
166 color (no diplopodia)
53 white and diplopodia

White is dominant to color and normal is dominant to diplopodia so:

180 white (no diplopodia)	*I_ D_*
68 color and diplopodia	*ii dd*
166 color (no diplopodia)	*ii D_*
53 white and diplopodia	*I_ dd*

Now consider each locus separately to simplify the problem. (Notice how often simplification comes from the approach of considering one locus separately.)

The feather color phenotypes occur in a ratio of about 1:1 (233 white: 234 color). That is characteristic of a test cross, *Ii* X *ii*.
The diplopodia phenotypes occur in a ratio of about 3:1 (346:121). That is characteristic of a cross of two heterozygotes where one allele is dominant, or *Dd* X *Dd*.
So the original cross was probably *Ii Dd* X *ii Dd*.

PRACTICE PROBLEMS.

8-8. A cross in chickens involved the diplopodia gene (See Problem 8-7) and "naked neck" (*n*), a recessive mutation at another locus that results in loss of feathers on the neck. The offspring included 85 normal, 10 with naked neck and diplopodia, 26 with naked neck and 30 with diplopodia. What were the genotypes and phenotypes of the parents?

8-9. Horses can have big white spots (a recessive allele, *b*) or little white spots (a dominant allele, *B*). They can also have weak hooves (*W*) or strong hooves (*w*), and normal nostrils (*F*) or flaring nostrils (*f*). A woman admired a horse named Izzy and wanted to buy a colt that had Izzy as a parent. Izzy's owner wanted to sell her Lucinda, a cute colt with big white spots, weak hooves and flaring nostrils. Izzy has little white spots, strong hooves and normal nostrils. Could Izzy be a parent of Lucinda?

8-10. Mutants of beans have been found that produce color variations in the seedlings. Two such mutations were crossed with the following results:

	Leaf Color			
Crosses	**normal green**	**lighter green**	**pale-green**	**yellow-green**
T X T	all			
S X S		all		
T X S (F_1)	59			
F_1 X F_1 (F_2)	311	107	128	32
F_1 X S	14	13		
F_1 X T	20		17	

Define some gene symbols and then show the genotypes of T, S, the F_1 and the F_2.

CHAPTER 3
MORE COMPLEX SITUATIONS

9. Many genetics problems go beyond these simplest situations due to complex relations between genotype and phenotype

A. No dominance

When there is dominance, the heterozygote has the same phenotype as the dominant homozygote, but dominance is not always observed. Many genes show incomplete dominance, in which the heterozygote has a phenotype intermediate to the two homozygotes. For example, a heterozygote with alleles for red flowers and for white flowers might be pink. Other genes show codominance, another case of no dominance, in which the heterozygote has the phenotype of both homozygotes. For example, a person heterozygous for the A and B blood types is type AB, having both the A and B blood cell antigens. **Incomplete dominance and codominance make your task a little easier because all of the genotypes have unambiguous phenotypes.**

The symbols used when there is no dominance are sometimes slightly different than for dominant alleles. Lowercase letters tend to be reserved for recessive alleles. Often when there is no dominance, each allele will be given a number or different letter as a superscript. For example: R^1, R^2, etc. or I^A, I^B, etc.

Solved Problem 9-1. If you cross true breeding four-o-clock plants with red flowers to true breeding four-o-clock plants with white flowers, the resulting heterozygotes have pink flowers. What will be the ratio of red, white and pink flowers if two pink flowered plants are crossed?

Answer. Let R^1R^1 be red and R^2R^2 be white. Pink will be R^1R^2.

P: R^1R^2 X R^1R^2

Gametes: half the gametes from each parent will be (R^1)

half the gametes from each parent will be (R^2)

Punnett Square:

	R^1	R^2
R^1	R^1R^1 Red	R^1R^2 Pink
R^2	R^1R^2 Pink	R^2R^2 White

So the ratio will be 1 red: 2 pink: 1 white. The difference from a cross in which red would be dominant is that the pink flowers can be distinguished from the red flowers.

Solved Problem 9-2. A true-breeding radish with long red roots was crossed to a true breeding radish with round white roots. The F_1 radishes were all oval and purple. Assuming that two independently assorting loci are involved, what would be the ratio of all the possible phenotypes among the F_2 produced by crossing two of the F_1 radishes?

Answer. Diagram the cross first.

P:	Long Red	X	Round White
F_1:		All Oval Purple	
F_2:		??	

We know that the parental radishes are all homozygous (true breeding) for two independently assorting loci, probably one for color and one for shape. The F_1 is probably heterozygous for both loci. We know that the F_1 is different from either parental type, so it is probably a case of incomplete dominance for both loci. Let's choose some gene symbols:

"Shape" locus	S^L = long	S^N = round
"Color" locus	C^R = red	C^W = white

(Because no alleles are dominant, we don't use the upper and lower case symbols.) Now diagram the cross again with all that we know.

P:	Long Red ($S^LS^L\ C^RC^R$)	X	Round White ($S^NS^N\ C^WC^W$)
F_1:		All Oval Purple ($S^LS^N\ C^RC^W$)	
F_2:		??	

Now we are ready to answer the question. Cross two F_1 plants, determine the kinds of gametes they will produce and construct a Punnett Square .

$S^LS^N\ C^RC^W$ will produce four kinds of gametes: S^LC^R S^LC^W S^NC^R S^NC^W So the Punnett Square will be:

	S^LC^R	S^LC^W	S^NC^R	S^NC^W
S^LC^R	$S^LS^LC^RC^R$ Long Red	$S^LS^LC^RC^W$ Long Purple	$S^LS^NC^RC^R$ Oval Red	$S^LS^NC^RC^W$ Oval Purple
S^LC^W	$S^LS^LC^RC^W$ Long Purple	$S^LS^LC^WC^W$ Long White	$S^LS^NC^RC^W$ Oval Purple	$S^LS^NC^WC^W$ Oval White
S^CC^R	$S^LS^NC^RC^R$ Oval Red	$S^LS^NC^RC^W$ Oval Purple	$S^NS^NC^RC^R$ Round Red	$S^NS^NC^RC^W$ Round Purple
S^CC^W	$S^LS^NC^RC^W$ Oval Purple	$S^LS^NC^WC^W$ Oval White	$S^NS^NC^RC^W$ Round Purple	$S^NS^NC^WC^W$ Round White

The phenotypic ratios can be found by counting up the number of each phenotype in the Punnett Square:

Long Red	1/16	Oval Red	2/16	Round Red	1/16
Long Purple	2/16	Oval Purple	4/16	Round Purple	2/16
Long White	1/16	Oval White	2/16	Round White	1/16

PRACTICE PROBLEMS

9-3. Decide whether each of the following descriptions is an example of dominance, incomplete dominance or co-dominance.

A. People with sickle cell anemia are homozygous for the *s* allele (*ss*). The homozygote *SS* does not suffer from sickle cell anemia. *Ss* heterozygotes are sometimes said to have "sickle cell trait." Under severe stress they may suffer some of the symptoms of sickle cell anemia, but to a much-reduced degree of severity.

B. Two plants of grain amaranth with black seeds were crossed to each other. Among the progeny, there 3 plants with black seeds for each plant with yellow seeds.

C. The D' allele in horses dilutes the resulting color of some other coat color genes. The homozygote $D'D'$ dilutes more than the heterozygote $D'D$.

D. Blood group M is due to the *M* allele in homozygous condition. Blood group N is due to the *N* allele in homozygous condition, and blood group MN is due to the *MN* heterozygote.

E. Tropical fish called "sword tails" are olive green in their natural habitat. A variant isolated by hobbyists is bright orange with black spots and is called "Montezuma." A single locus is involved, but a pure breeding strain of Montezuma has not been produced because homozygous Montezuma is lethal -- it results in dead fish. Heterozygotes are "Montezuma" and homozygous wild type are olive green.

9-4. Would the results of Problem 9-2 have been any different if we had started with true breeding radishes that were long white and round red?

9-5. In cattle the coat can be either red (C^RC^R), white (C^WC^W) or roan (C^RC^W). Roans have both white and red hairs. Horns are either normal (*hh*) or missing (*HH* or *Hh*). The coat color alleles show codominance, and the horn alleles show dominance. Suppose we crossed two C^RC^W *Hh* cattle. What proportion of their offspring would have horns and be roan? What proportion will lack horns and be white?

B. Multiple alleles

So far we have limited ourselves to loci with only two alleles, winged vs. wingless, striped vs. solid, disease vs. no disease. But there can be more than two alternate forms of one locus. A single locus represented by 3 or more alleles in a population is said to have multiple alleles. Solving genetics problems with multiple alleles is not that much more difficult than solving simpler problems since diploid organisms can still only have two alleles at any locus. The important thing is to be systematic. As long as you consider what is happening in only one individual at a time, every multiple allele problem becomes a collection of problems involving two alleles.

Solved Problem 9-6. It was first thought that the albino locus in rabbits had only two alleles, the wild type *C* and the recessive *c*. Recessive homozygotes lack melanin and have white hair and pink eyes. Several other alleles of this locus are now known. A slightly more complicated symbol system is used that adds a superscript to alleles. In addition to the dominant *C* and recessive *c* allele designations, the "chinchilla" c^{ch} and "Himalayan" c^h are used to unambiguously designate the multiple alleles for this locus. Any single rabbit inherits only two of these alleles, which segregate in a normal fashion during meiosis. Write the genotypes of the progeny that will result from each of the following crosses:

A) Cc X c^hc^h B) Cc X c^hc^{ch} C) cc X c^hc^{ch}

Answer.

A) P: Cc X c^hc^h

Gametes: 1/2 (C) 1/2 (c) all (c^h)

Punnett Square

	C	c
c^h	$C\,c^h$	$c\,c^h$

B) P: Cc X c^hc^{ch}

Gametes: 1/2 (C) 1/2 (c) 1/2 (c^h) 1/2 (c^{ch})

Punnett Square

	C	c
c^h	$C\,c^h$	$c\,c^h$
c^{ch}	$C\,c^{ch}$	$c\,c^{ch}$

C) P: cc X c^hc^{ch}

Gametes: all (c) 1/2 (c^h) 1/2 (c^{ch})

Punnett Square

	c
c^h	$c\,c^h$
c^{ch}	$c\,c^{ch}$

In problem 9-6 we only looked at the genotypes. When considering phenotypes in multiple allele problems, dominance is a little more complicated. "Dominant" and "recessive" are relative terms that refer to what happens in a heterozygote -- if there is dominance, the heterozygote will look just like the dominant homozygote. When there are multiple alleles, an allele may be dominant to one allele and recessive to another or show incomplete or codominance with one allele and complete dominance with another. Multiple allele problems are much easier if you first get as much genotype information figured out as you can and then work on the phenotypes.

Solved Problem 9-7 The human ABO blood group locus has three alleles, I^A, I^B and i^o. I^A and I^B are both dominant to i^o, but they show codominance in I^AI^B heterozygotes. The ABO alleles control antigens on the surface of red blood cells that can be recognized with the appropriate antibodies. Persons whose cells have A antigen are said to be Type A, those with B antigen are Type B, those with both the A and B antigens are Type AB, and those without either antigen are Type O. Here is a chart of the relation of genotypes to phenotypes.

Blood Group	Genotypes
A	$I^A I^A$ or $I^A i^o$
B	$I^B I^B$ or $I^B i^o$
AB	$I^A I^B$
O	$i^o i^o$

What are the expected phenotypes produced among the offspring of two people who are type AB? Among the offspring of a type A heterozygote and a Type B heterozygote?

Answer. First lay out what is known about the genotypes, work that out for parents and offspring and then deal with phenotypes.

Type AB people have the unambiguous genotype $I^A I^B$. The cross looks like this and is like any other cross we've considered with alleles lacking dominance.

P: $I^A I^B$ X $I^A I^B$

Gametes: (I^A) and (I^B) (I^A) and (I^B)

Punnett Square (Fill in the genotypes and <u>then</u> worry about the phenotypes.)

First do genotypes

	I^A	I^B
I^A	$I^A I^A$	$I^A I^B$
I^B	$I^A I^B$	$I^B I^B$

Then add phenotypes

	I^A	I^B
I^A	$I^A I^A$ Type A	$I^A I^B$ Type AB
I^B	$I^A I^B$ Type AB	$I^B I^B$ Type B

In the second cross, all three alleles are involved, but always only two are in any one individual.

P: I^Ai^o X I^Bi^o

Gametes: (I^A) and (i^o) (I^B) and (i^o)

Punnett Square (Fill in the genotypes and <u>then</u> worry about the phenotypes.)

First do genotypes

	I^A	i^o
I^B	I^AI^B	I^Bi^o
i^o	I^Ai^o	i^oi^o

Then add phenotypes

	I^A	i^o
I^B	I^AI^B Type AB	I^Bi^o Type B
i^o	I^Ai^o Type A	i^oi^o Type O

Solved Problem 9-8. Mrs. Idengaku was one of two mothers in a maternity ward. When she was given baby #1, she denied that it was hers, claiming baby #2 instead. The other mother also claimed baby #2.

Mrs. Idengaku is blood type O. Baby #1 is A, and Baby #2 is O. Unfortunately, Mr. Idengaku died just before the baby was born, so we can't find out his phenotype or genotype, but the Idengakus had 3 other children whose phenotypes are known. Keiko is A, Tohru is B and Kenichi is B. Who is right, Mrs. Idengaku or the other mother? What is your reasoning?

Answer: Set up the cross between Mr. and Mrs. Idengaku and put in all the information we have from the problem.

Mrs. Idengaku	Mr. Idengaku
Group O	Unknown blood group
i^oi^o	_ _

Keiko	Tohru	Kenichi
Group A	Group B	Group B
I^A _	I^B _	I^B _

Baby #1	Baby #2
Group A	Group O
I^A _	i^oi^o

There are Idengaku children with both I^A and I^B alleles. Since Mrs. Idengaku is i^oi^o, those alleles must both have come from the father. So Mr. Idengaku was I^AI^B. If he was I^AI^B, the Idengakus cannot have had a child that is i^oi^o, since Mr. Idengaku has no i^o allele to give the baby. They could, however, have a child that is Group A. Their group A children would be $I^A i^o$. Thus Mrs. Idengaku is mistaken -- Baby #2 cannot have been an Idengaku. Baby #1 could be, however.

PRACTICE PROBLEMS

9-9. With regard to ABO blood groups, one parent in a family is group A and the other is group B. If all four blood groups appear among the offspring, what proportion of the offspring will be group O?

9-10. The rabbit coat color genes described in Problem 9-5 show a complex form of stepwise dominance. When the c (albino) allele is homozygous, it prevents pigment formation, so the homozygous individual has white hair. Allele c^h produces "Himalayan" rabbits with light bodies and dark noses. It is dominant to albino. The c^{ch} (chinchilla) allele produces a coat color that is lighter than wild type but darker than Himalayan, and it produces a "light gray" coat color when heterozygous with c or c^h. Allele C is dominant to all three of the other alleles. We can represent that as a dominant series $C > c^{ch} > c^h > c$. What are all the types of coat color that will be seen in the parents and offspring of a cross between a Cc^{ch} heterozygote and a $c^h c$ heterozygote?

9-11. A multiple allele series in domestic cattle influences the spotting pattern of the coat. Allele S produces "Dutch belt," s^h the Hereford pattern, s^c a solid color and s the Holstein pattern. These genes show an order of dominance in which S is dominant to the other three, s^h is dominant to all but S, s^c is dominant only to s, and s is recessive to all the other alleles (i.e., $S > s^h > s^c > s$).

A) Can Holsteins produce Hereford offspring?

B) Can Herefords produce Holstein offspring?

C) Can both solid color and Holstein cattle be among the offspring of a single pair of Dutch belt cattle?

C. Gene interactions: epistasis

Sometimes alleles at one locus will have an effect on the expression of alleles at another locus. When a gene at one locus masks the expression of genes at another locus, the process is called epistasis. The result of epistasis is a modification of the phenotype ratios you would expect if there were no epistasis. The genes still assort independently, but some of the phenotypes may be reduced in number or missing. You do these problems exactly as you would any other problem involving two or more genes. When you come to determining the phenotypes, however, you have to be more careful since you have to take those interactions into account.

Solved Problem 9-12. Black mice and brown mice differ in how the pigment melanin is distributed in the hair. Allele *B* is dominant and results in black hair, while allele *b* is recessive and results in brown hair when homozygous. Another locus determines whether the pigment melanin is synthesized at all. The recessive *c* allele blocks melanin production so the resulting "albino" mice have white hair. The dominant *C* permits melanin production. Obviously, if there is no melanin production due to *c*, it does not matter whether *B* or *b* is present. So allele *c* is epistatic to the *B/b* alleles when *c* is homozygous. What would be the phenotypes and genotypes of parents and their offspring if we crossed two mice heterozygous for both loci?

Answer: Diagram what we know already from the problem.

Bb Cc X Bb Cc

These mice have a *C* allele, so they can make pigment, and they have a *B* allele, so they will be black.

Each parent can produce four kinds of gametes: (BC) (Bc) (bC) and (bc.) So a Punnett Square will look like this:

	BC	Bc	bC	bc
BC	BB CC Black	BB Cc Black	Bb CC Black	BbCc Black
Bc	BB Cc Black	BBcc Albino	BbCc Black	Bb cc Albino
bC	BbCC Black	BbCc Black	bb CC Brown	bbCc Brown
bc	Bb Cc Black	Bb cc Albino	bb Cc Brown	bb cc Albino

If this cross of two heterozygotes had involved two loci, each with dominance but with no epistasis, we know that we should expect a 9:3:3:1 ratio of phenotypes among the progeny. But epistasis by the albino allele when homozygous masks the color gene. So:

B_ C_	Black	9/16	bb C_	brown	3/16
B_ cc	white	3/16	bb cc	white	1/16

That is 9 black: 3 brown: 4 white. We see the same gene arrangement as always, but different phenotype interactions. The easiest way to do these problems is first to work out the segregation and independent assortment of genes and then to determine the phenotypes.

Solved Problem 9-13. A gray horse was crossed to a chestnut horse. All the offspring were gray. Then two of the F_1 were crossed. Among the F_2, 12/16 were gray, 3/16 were bay, and 1/16 were chestnut. Explain these results by showing the genotypes of each phenotype among the parents, the F_1 and the F_2.

Answer. This problem is the reverse of Problem 9-10. This time you are being given phenotype data from which you have to figure out genotypes. In 9-10 you were given the genotypes and how the genes interacted and were told to figure out phenotypes. The secret of this second kind of problem is to find a ratio that you recognize and build on that.

Diagram what you know:

P:	Gray X Chestnut		
F_1	All Gray		
F_2	12/16 Gray	3/16 Bay	1/16 Chestnut

Whenever you see sixteenths, you should think of the 9/16 : 3/16 : 3/16 : 1/16 ratio that results from crossing two organisms that are heterozygous for two independently assorting alleles. In this case the 9:3:3:1 ratio appears to have been modified by some sort of epistasis. One of the 3/16 classes has been combined with the 9/16.

F_1	Aa Bb X Aa Bb			
F_2	9/16 A_ B_	3/16 A_ bb	3/16 aa B_	1/16 aa bb

The 1/16 category is unambiguously the double homozygous recessive. So chestnut has that genotype. It looks as if the *A* allele masks the expression of the *B* and *b* allele, so that whenever there is an *A* allele, the phenotype is gray:

9/16 A_ B_ (Gray); 3/16 A_ bb (Gray); 3/16 aa B_ (Bay); 1/16 aa bb (Chestnut)

And then, of course, the *aa B_* genotype produces the bay phenotype. In problem 9-10, the epistatic allele (albino) was recessive. Here, the epistatic allele (gray) is dominant so we get a different modification of the 9:3:3:1 ratio.

One parent was chestnut, so it must have been *aa bb*. To produce the double heterozygous F_1, the other parent must have been *AA BB*. We would expect that genotype to be gray, and it is!

PRACTICE PROBLEMS

9-14. A 9:3:3:1 ratio occurs when two double heterozygotes are crossed. Gene interactions modify the 9:3:3:1 ratio. If a double heterozygote is test crossed (crossed to a double homozygous recessive), a 1:1:1:1 ratio is obtained when there is no gene interaction, but epistasis will alter this 1:1:1:1 ratio. For each of the following modified 9:3:3:1 ratios, what will be the modified 1:1:1:1 ratio if the double heterozygote is test crossed?

a) 12:3:1 b) 9:7 c) 15:1

9-15. Because of the medical importance of mosquitoes, their genetics has been studied carefully, and many different genes have been found. In one species thc larvae are ordinarily tan with a light colored mid-dorsal stripe running along the abdomen. The allele *s* eliminates the stripe when homozygous. The *S* allele is dominant and results in the stripe. Another locus controls the body color, with the dominant allele *B* resulting in tan and the recessive *b* resulting in dark brown when homozygous. Allele *b* also reduces the mid-dorsal stripe to a tiny dot on the third segment of the abdomen. Suppose a homozygous normal mosquito is crossed to a dark brown mosquito with neither stripe nor spot. What will be the phenotypes of the F_1 and F_2 generations?

9-16. A black mouse was crossed to a hairless mouse. Since it lacked hair, it was impossible to tell if its hair was black or brown! All the progeny had black hair. When these F_1 progeny were mated among themselves, 100 F_2 were black, 43 were hairless and 30 were brown. What was the genotype of the original parents?

9-17. Geneticists who wanted to learn more about cucumber genetics began to search for genetic variation in the plants. They found genes for the color of the spines on the fruit. The spines can be either black or white, and the inheritance pattern they observed was most simply explained as due to two independently assorting loci. At least one dominant allele had to be present at each locus for spines to be black. Otherwise spines were white. If they crossed two double heterozygotes and obtained 1000 offspring, how many plants would have white-spined fruit, and how many would have black?

CHAPTER 4
SEX LINKAGE

10. Sex-linkage

Some organisms have a chromosomal method of sex determination in which the two sexes have differences in one pair of chromosomes. In humans and fruit flies, females and males differ in a single pair of chromosomes called the sex chromosomes. In the females the two sex chromosomes are both X-chromosomes, while in the males there is an X-chromosome and a Y-chromosome. Often the Y-chromosome contains few genes while the X-chromosome contains many genes, so there are loci on the X that are not on the Y. The genes on the X that are not on the Y are called X-linked genes. (Sometimes they are called "sex-linked" genes, but, strictly speaking, sex linkage includes rare cases of Y-linked genes.) X-linked genes add a complication to genetics problems. All the other chromosomes that are not sex chromosomes are called autosomes. All the problems in this book up to this point have involved genes on the autosomes, called autosomal genes.

A. Use informative symbols

Problems involving X-linked genes are not difficult to do if you remember that the X and Y chromosomes segregate much as any pair of alleles would segregate and if you use a system of symbols to remind you of sex linkage. A simple way to do this is to write an X or a Y for the chromosome and then write the alleles being studied as if connected to the X. For example:

A female fly heterozygous for the white eye gene (*w*): X^WX^w

A male fly with the white eye gene: X^wY

A female human heterozygous for muscular dystrophy: X^DX^d

A male human with the allele for red-green color blindness: X^cY

Notice that females can be either heterozygous or homozygous, but males have only one copy of the gene, so those terms can't be applied. Instead we say that males are hemizygous.

Solved Problem 10-1. Define symbols and write the genotypes of each of the following individuals.

A) A female fruit fly heterozygous for the X-linked gene for forked bristles

B) A female human heterozygous for the recessive hemophilia gene.

C) A female human homozygous for the hemophilia gene.

D) A male human with the muscular dystrophy gene.

E) A male human without the allele for red-green color blindness.

Answer: First write the sex chromosome composition of each individual and then hook on the appropriate allele symbol.

A) A female, so XX. Let *f* be forked bristles. $X^F X^f$

B) A female, so XX. Let *H* be normal and *h* be hemophilia. $X^H X^h$

C) A female, so XX. $X^h X^h$

D) A male, so XY. Let the allele for muscular dystrophy be *d*. X^dY

E) A male, so XY. Let *c* be the color blindness allele and *C* its normal allele. X^CY

B. Keep the alleles attached to their chromosomes.

The X and Y chromosomes segregate into gametes much as a pair of alleles would. Half the sperm will receive a Y chromosome and half will receive an X. All the eggs will receive an X. If the female is heterozygous at an X-linked locus, half of her eggs will contain an X with one of the alleles on it, and half will contain an X with the other allele on it. Remember that, and you can't go wrong.

Solved Problem 10-2. The first allele discovered by *Drosophila* geneticists was the recessive white eye allele *w*, which is X-linked. The normal eye is red. A red eyed male was crossed to a white-eyed female, and a white-eyed male was crossed to a red eyed female. All parents are from true breeding lines. What kinds of progeny will be found in each case?

Answer. Define symbols and diagram each cross. Let X^W be the symbol for a chromosome bearing the red-eyed allele and X^w the white-eyed allele. Remember that the Y has neither allele, so it is just Y.

P:	Red male X^WY	X	White female $X^w X^w$
Gametes:	1/2 X^W		all X^w
	1/2 Y		

	White male X^wY	X	Red female $X^W X^W$
Gametes:	1/2 X^w		all X^W
	1/2 Y		

Punnett squares:

	X^W	Y
X^w	$X^W X^w$ Red Female	$X^w Y$ White Male

	X^w	Y
X^W	$X^W X^w$ Red Female	$X^W Y$ Red Male

Solved Problem 10.3. In *Drosophila* forked bristles (*f*) is an X-linked character. Straight bristles (*F*) is dominant. Vestigial wings (*v*) is autosomal recessive, and its wild type allele (*V*) is dominant. A male with normal bristles and homozygous for normal wings is crossed to a vestigial winged female that is heterozygous at the forked locus. Show the genotypes and phenotypes of their offspring.

Answer. Outline the cross.

A) P:	VV X^FY	X	vv $X^F X^f$
	Male Normal Wings Normal Bristles		Female Vestigial Wings Normal Bristles
Gametes:	1/2 VX^F 1/2 VY		1/2 vX^F 1/2 vX^f

Punnett Square:

	V X^F	VY
v X^F	Vv X^FX^F Female Normal Wings Normal Bristles	Vv X^FY Male Normal Wings Normal Bristles
v X^f	Vv X^FX^f Female Normal Wings Normal Bristles	Vv X^fY Male Normal Wings Forked Bristles

All the females and half the males have normal wings and normal bristles. Half the males have normal wings and forked bristles.

PRACTICE PROBLEMS

10-4. Red-green color blindness is a recessive X-linked character in humans. In cases involving two parents with normal color vision, can male offspring be color blind? Can female offspring?

10-5. Look at the results of Problem 10-3. If one of the male offspring with forked bristles is crossed to a female heterozygous for both loci, how many of each possible type would you expect if there were 400 progeny from that cross?

10-6. Hemophilia is a defect in blood clotting due to a recessive sex-linked allele. Tzarevich Alexis, son of the last Russian Tzar and Tzarina, Nicholas II and Alexandra, had hemophilia. Rupert Viscount Trematon of England, son of Alice of Athlone, also had hemophilia. The royal families of England and Russia were related. From this information is it more probable that Alice was a cousin of Nicholas II or of Alexandra? Explain.

3. You can sometimes recognize the patterns of sex-linked inheritance.

As with the autosomes, some problems tell you about the parents and ask you about the offspring, but some problems give you information about the offspring and expect you to recognize that the pattern of inheritance reveals sex-linkage. Several clues suggest sex-linkage. Not all of these clues will be found in every case, but any of them should cause you to consider sex-linkage as a possibility.

Clue 1. Among the progeny, the male phenotypes may be in different ratios than the female progeny. Look, for example at the cross in Problem 10-3. The bristle gene is X-linked, and the vestigial gene is not. Both the male and female progeny have normal wings, but some of the male progeny have forked bristles while all of the female progeny have normal bristles. On the other hand, notice Practice Problem 10-5. The males and females have the same phenotype for both wing and bristles. The presence of one of these clues is good evidence for sex-linkage, but the absence of clues does not eliminate the possibility of X-linkage. Sometimes you have to look at several different crosses to be sure.

Clue 2. Reciprocal crosses may give different results. In reciprocal crosses, the phenotype of the male in one cross is the phenotype of the female in the other. Problem 10-2 involves reciprocal crosses of the white eye gene. Notice that in one cross all the offspring have red eyes, but in the reciprocal cross the males have white eyes.

Solved Problem 10-7. Suppose a recessive allele *l* causes death in the early embryo so that homozygotes for *l* are never counted among the progeny. What would be the effect if the gene were autosomal and a female heterozygous for *l* were crossed to a normal homozygous male? What if the gene were X-linked and a heterozygous female were crossed to a normal male?

Answer. Outline the two crosses.

	If Autosomal	If X-linked
P:	LL X Ll Male Female	X^LY X X^LX^l Male Female
Gametes:	all (L) 1/2 (L) 1/2 (l)	1/2 (X^L) 1/2 (Y) 1/2 (X^L) 1/2 (X^l)
F_1	1/2 LL; 1/2 Ll All normal No difference in the two sexes	1/4 X^LX^L; 1/4 X^LX^l Females all normal 1/4 X^LY; 1/4 X^lY (dead) Half the males missing

So when this lethal gene is X-linked, there will be half as many male progeny as female progeny. There is no effect on the sex ratio when the gene is autosomal.

Solved Problem 10-8. Red-green color blindness is a defect in the color sensitivity of the retina. It was noticed that it tended to appear in males more frequently than in females and that a male with red-green color blindness could have two normal parents. Usually men with red-green color blindness do not have color blind children, but they can have color blind grandchildren. What kinds of children and grandchildren would you expect in a rare case of a woman with red green color blindness, assuming that she married a man without the condition?

Answer. That two normal people can have a child with red-green color blindness is evidence that the gene is recessive. That red-green color blind men would have grandchildren with the problem but not children suggests sex-linkage.

Father is X^cY. All his daughters will get the X^c, but his sons will get the Y. If the mother is normal, none of their sons or daughters will be color blind. But grandsons that receive the X^c will be.

If red-green color blindness is recessive a woman with it, must be homozygous $X^c X^c$.

Among her children, all the males will get an X^c and will be color blind. All her female children will get the normal X^C from the father and will not be color blind.

Her sons are X^cY. If they have children by normal wives, none of the grandchildren will be color blind since all get an X^C from their mother. All her daughters, however, will be $X^C X^c$, so half her grandsons are expected to be color blind.

PRACTICE PROBLEMS

10-9. The Bar eye allele in *Drosophila* results in a reduction in the number of units in the compound eye so that the eye is long and narrow instead of round. This allele is dominant to the allele for normally shaped eyes. When males from a true-breeding Bar-eyed strain are crossed to normal females, the female offspring have Bar eyes, but the males do not. When normal males are crossed to females from a true-breeding bar-eyed strain, all the progeny have Bar eyes. Explain this result.

10-10. As indicated in Problem 10-2, white eye in *Drosophila* is an X-linked allele. A geneticist wondered if another gene (for body color) was X-linked also. She crossed black-bodied males to white-eyed females. From the results below, can you decide if the body color gene is X-linked? Explain.

P: red-eyed, black-bodied male X white-eyed, yellow bodied female

F_1:
males: white-eyed, yellow bodied
females: red-eyed, yellow-bodied

F_2:

males:		females:	
red yellow	3/16	red yellow	3/16
white yellow	3/16	white yellow	3/16
red black	1/16	red black	1/16
white black	1/16	white black	1/16

10-11. Here is a cross involving three different independently assorting loci. One of the loci is X-linked. Which one?

P: Long, Purple, Active, Male X Short, Yellow, Sluggish, Female

F_1:
1/4 Long, Yellow, Active, Males
1/4 Short, Yellow, Active, Males
1/4 Long, Purple, Active, Females
1/4 Short, Purple, Active, Females

CHAPTER 5
PEDIGREES

11. Pedigrees are another way to analyze inheritance (and they give you a chance to review the basic principles of solving genetics problems).

Humans present a special problem to geneticists because controlled crosses are not possible (or desirable). The numbers of offspring are often small, so working with ratios is not always possible either. As a systematic way of making the most of what data are available, geneticists use pedigrees, charts of several generations of inheritance. In pedigrees you examine past "crosses" looking at the pattern of appearance of phenotypes like those of the individual under study.

A. Individuals are symbolized by circles and squares.

The pedigree focuses on the traits of individual people. Males are symbolized with a square, females with a circle, and people of unknown gender with a diamond shape. People who have the particular phenotype being studied have their circle or square filled in. People without that phenotype have open circles or squares. A person who was still-born or aborted and thus not available to determine the phenotype is represented with a small dot.

Male　　Female　　Unknown　　Has the phenotype　　Aborted or stillborn

Solved Problem 11-1. Suppose we were studying the condition called cystic fibrosis. Write the symbol for a man with cystic fibrosis and for a woman without it.

Answer: Man with cystic fibrosis: ■ Woman without cystic fibrosis: ○

B. Relationships between individuals are represented by lines.

A mating between two people is symbolized by drawing a horizontal line to connect them. Offspring of a mating are connected by a vertical line below the line between their parents and to each other by a horizontal bracket. Usually progeny are represented from left to right in order of their birth. For example, here is a mating that produced a daughter first and then a son.

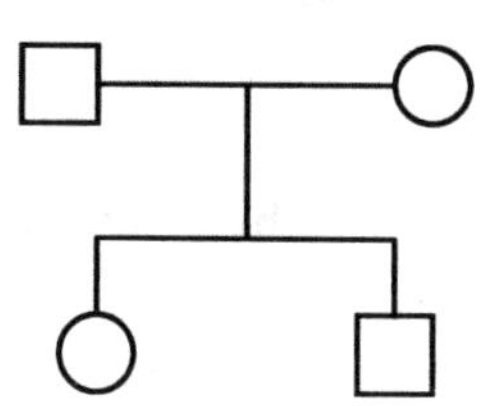

Solved Problem 11-2. A man marries and has two sons and a daughter. Then his wife dies and he marries again, having one daughter by the second marriage. Draw a pedigree that illustrates that. (Hint: Connect the man to both of his wives.)

Answer:

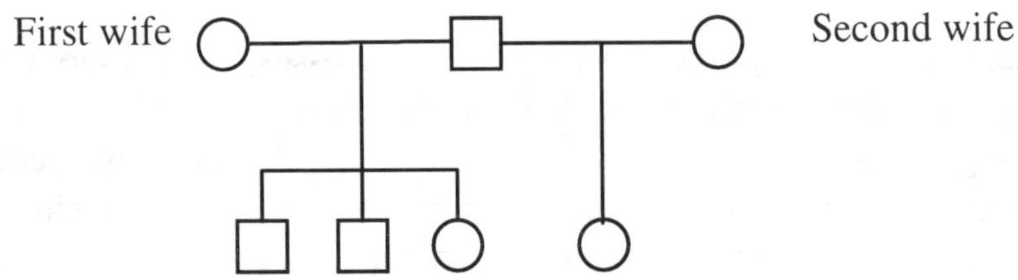

PRACTICE PROBLEMS

11-3. Two normal people gave birth to two girls who were normal and then to a boy who had Tay-Sachs disease. Draw that pedigree.

11-4. Lionel and Leona met at a singles dance and married soon afterward. They were initially attracted to each other because they both had "free" ear lobes, that is, earlobes not directly attached to their heads. After a respectable interval, they had a child, which to their surprise had "attached" ear lobes. Draw a pedigree to show Lionel, Leona and their child, whose name is Louella.

11-5. In a study of polydactyly, two people who had the condition produced a male child who also had the condition. That child eventually married and his wife (who did not have the condition) gave birth to a girl without the condition. Draw the pedigree that shows that.

C. Pedigrees can represent the genetic history of many generations.

As Problem 11.5 shows, it is possible to include any number of generations on a pedigree and to show matings between any number of people on each generation. When more than two generations is studied, usually each generation is numbered with a Roman numeral and each person at any one generation is numbered with an Arabic numeral. That makes it easier to refer to an individual.

Solved Problem 11-6. A pedigree follows several generations of a family afflicted with a hereditary disability. Those who have the disability are symbolized with a filled in symbol. Answer the following questions using the reference numbers for pedigrees.

a) Which individuals have the disability? b) Who are the grandchildren of III-4?
c) Which married couple are second cousins (i.e. the children of cousins)?

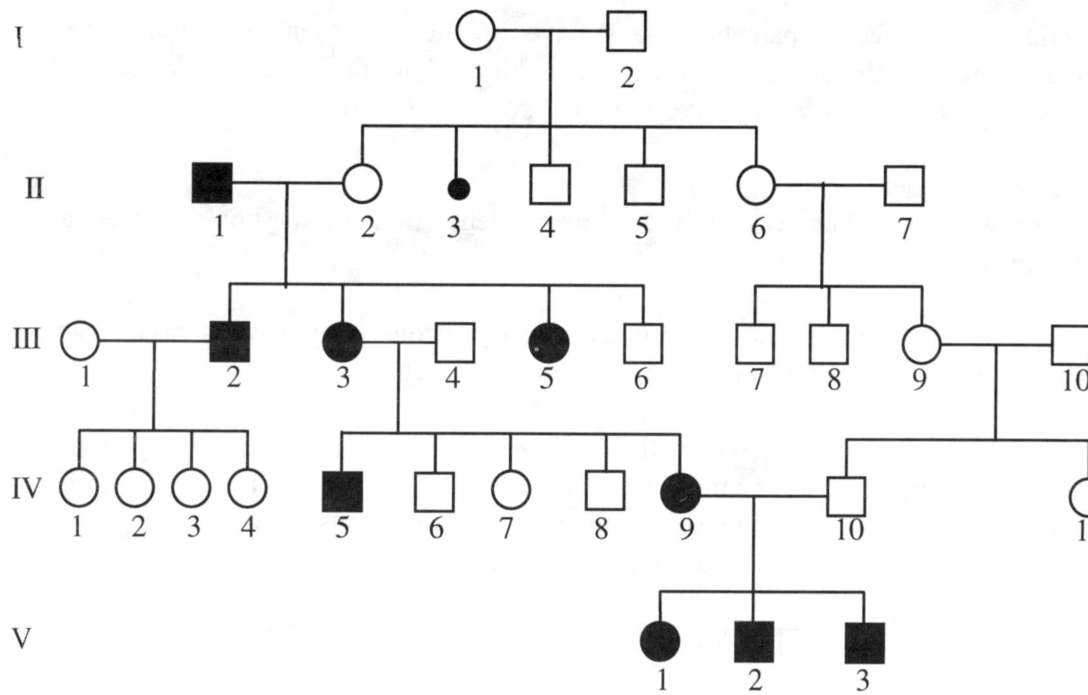

Answer. a) All those with filled symbols have the disability -- II-1, III-2, III-3, III-5, IV-5, IV-9, V-1, V-2, V-3.
b) Grandchildren are two generations after grandparents, so V-1, V-2 and V-3 are the ones. c) IV-9 and IV-10.

D. Analyze pedigrees using the same logic used for crosses.

Pedigrees represent crosses, although they are not controlled crosses. So we can use the same reasoning to learn from pedigrees that we used to learn from crosses. A pedigree represents a description of phenotypes, so pedigrees are problems where you are given phenotypes and you determine information about genotypes. Usually pedigree problems ask, "What kind of inheritance pattern does this gene show?" i.e. given the evidence on the pedigree, is the gene is dominant or recessive, and autosomal or sex-linked.

Here are some principles for analyzing pedigrees.

1. Don't take ratios too seriously in pedigrees. The numbers of individuals is often small, so if you have four offspring, a three to one ratio doesn't mean much. Sometimes you'll only have one offspring, and then ratios mean nothing at all. Instead, use the logic of sections 3 - 6 (Pages 4-17).

2. Is the condition of interest dominant or recessive? Several "Rules of Thumb" will help you make this decsion in a pedigree:

   If the condition is recessive, then

   1) expression of the trait often skips generations (the recessive allele is passed on heterozygotes and is not expressed unless two heterozygotes mate to produce a homozygous recessive child).

2) without exception, two affected parents always have affected children (two homozygous recessives can give rise only to homozygous recessive children); but unaffected parents may have affected children (two heterozygotes may have a homozygous recessive child).

If the condition is dominant, then

1) the condition shows up frequently, often in every generation (both homozygotes and heterozygotes express the phenotype).

2) affected parents may have unaffected children (two heterozygotes may have a homozygous recessive child); but unaffected parents never have affected children.

Solved Problem 11-7. The skin of individuals with *xeroderma pigmentosum* (a hereditary disease) is extremely sensitive to sunlight, often becoming thickened, freckled and cancerous. Based on the following hypothetical pedigree chart, can you say whether the disease is caused by a dominant or recessive allele?

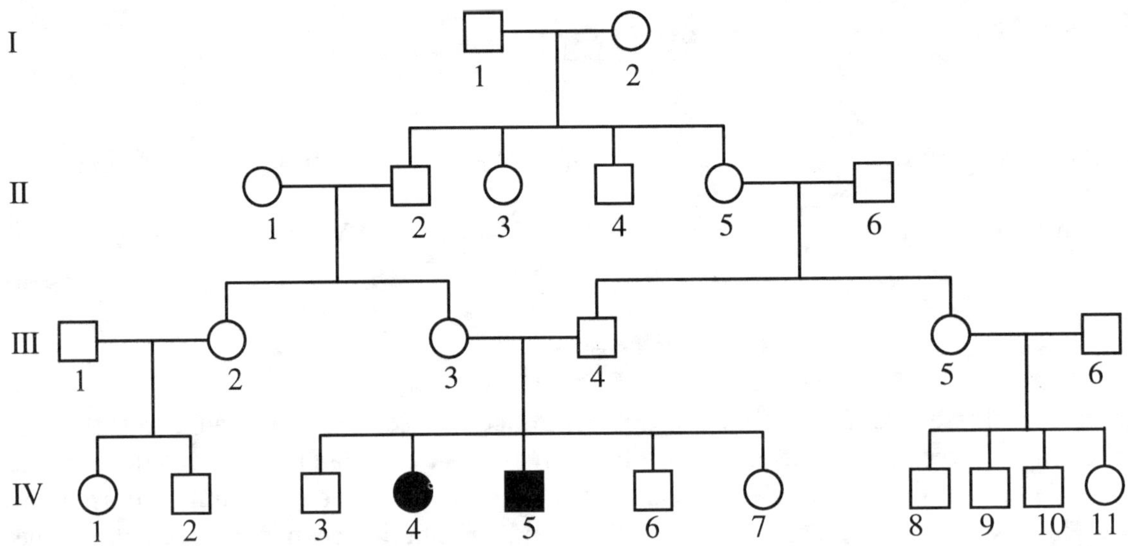

Answer: The condition must be recessive – it skips generations and appears among the offspring of unaffected parents, who must both be heterozygotes. (Note that these heterozygous parents are cousins -- xeroderma pigmentosum is rare.)

Solved Problem 11-8. Albinism is an absence of pigmentation determined by an autosomal recessive allele. Which of the children of the following crosses would be affected by this trait? (A) Albino parents, five children: girl, girl, boy, girl, boy. (B) Albino mother and homozygous non-albino father, five children: boy, boy, girl, girl, girl. (C) Homozygous non-albino parents, five children: boy, girl, boy, boy, girl. Draw pedigrees to illustrate each of your answers.

Answer. (A) Both parents are homozygous recessive, and therefore their children will be the same – homozygous recessive. Because albinism is a recessive condition, the whole family is affected, and the pedigree looks like this:

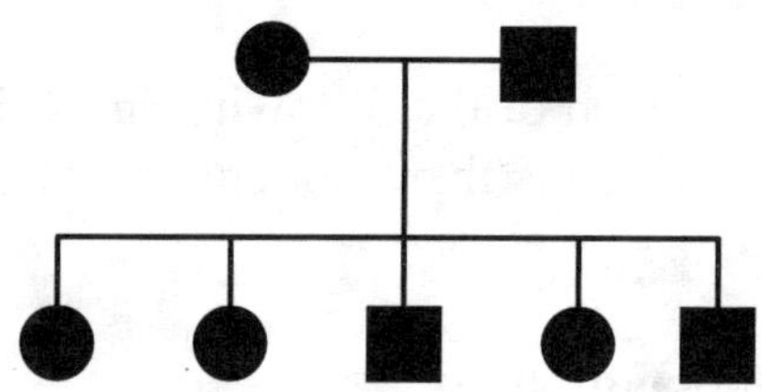

(B) The parents are homozygous for different alleles, and therefore their children will be heterozygous – unaffected, but carriers of an allele for albinism.

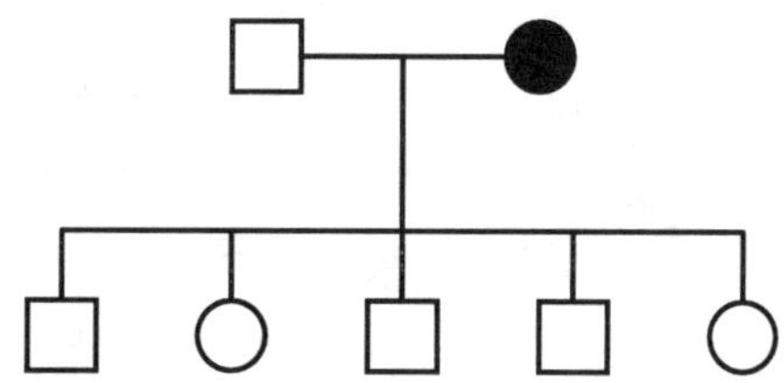

(C) Both parents are homozygous dominant, and therefore their children will be homozygous dominant also. The trait won't be expressed or even carried by any member of this family.

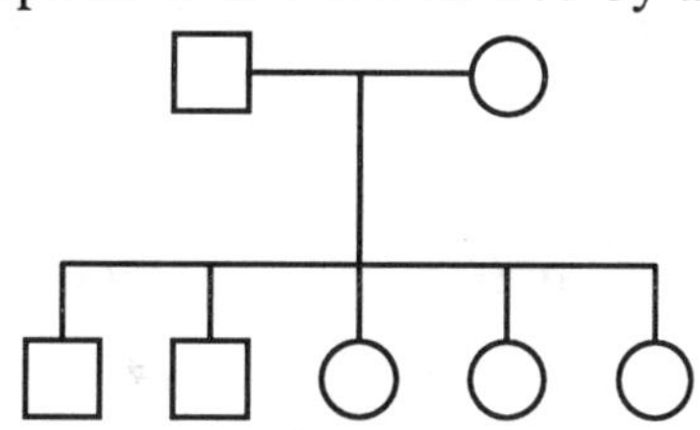

PRACTICE PROBLEMS

11-9. Look at the pedigree in Problem 11-4. Is free ear lobes dominant or recessive to attached ear lobes?

11-10. Genetic *pituitary dwarfism* is a recessive condition. Here is a pedigree in which a man's deceased first wife (the crossed out circle) was not affected by pituitary dwarfism, but his second wife is. Is the man heterozygous for this trait or homozygous dominant? Explain why.

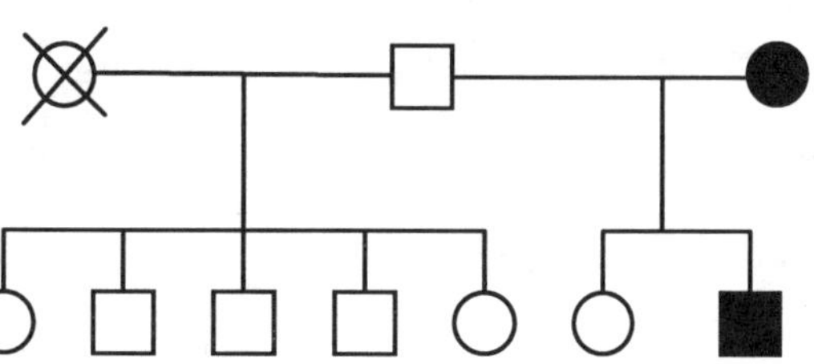

11-11. Look at the pedigree in Problem 11-5. Can you tell if polydactyly is dominant or recessive? Explain. Suppose the couple in the second generation had another daughter who had polydactyly. What would that tell you about dominance?

11-12. Look at the pedigree in Problem 11-6. Is this disability due to a dominant or recessive allele? Explain why.

11-13. Again look at Problem 11-6. Which of the following individuals is probably heterozygous for the disability and for which is there not enough evidence to decide? II-1, III-3, III-5, IV-9, V-2.

3. Is the character being studied autosomal or sex-linked?

Again, based on what you know about sex-linkage, there are some rules of thumb about how to tell if a gene is sex-linked. Assuming that sex-linked genes are only on the X and not on the Y and that males are XY and females XX,

If the gene is dominant,

1) a father will pass the allele to all of his daughters and none of his sons, and everyone who receives a copy will have the associated phenotype.

2) a female with a dominant X-linked allele will pass the allele to half of her offspring regardless of sex.

If the gene of interest is a recessive,

1) any male who has the gene will express it, since he is hemizygous.

2) recessive sex-linked phenotypes are more common among males than among females.

3) a man who has the gene will pass it on to all of his daughters and none of his sons. Since the gene is recessive, usually no offspring will express the gene.

4) but in the <u>next</u> generation the daughters will pass the gene to half of their sons. So a pattern in which the trait appears in the males of alternating generations may indicate sex-linkage.

Solved Problem 11-14. On page 47 is a pedigree in which the phenotype of interest appears in all the shaded individuals. Is the allele for the phenotype of interest dominant or recessive, and sex-linked or autosomal? Explain.

Answer: Children of unaffected parents are affected, so the allele must be recessive. It only shows up in males, and an affected male (II-8) does not pass the gene to his offspring, so the allele is probably sex-linked.

Solved Problem 11-15. On page 48 is a pedigree for yet another condition. Is the allele for the shaded phenotype dominant or recessive, and sex-linked or autosomal?

Answer: The phenotype only shows up when a parent has the gene, so the responsible allele is probably dominant. It would be nice to have one mating of two affected individuals to see if any unaffected offspring result -- as would be expected if the gene is dominant. Females who are affected pass it on to either males or females, but males only pass it on to females. So the gene is probably sex-linked.

PRACTICE PROBLEMS

11-16. Refer to Problem 11-14. Is I-1 homozygous or heterozygous at the locus in question? How do you know?

11-17. Look at the accompanying pedigree. Is the condition due to a dominant or recessive allele? Sex-linked or autosomal? How do you know?

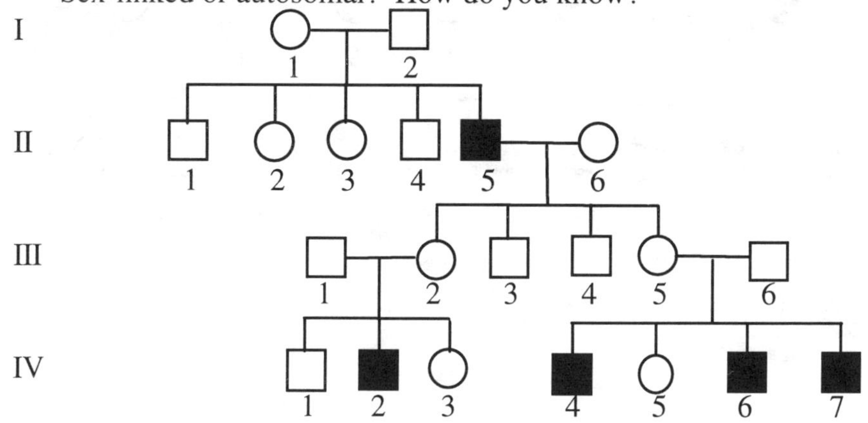

11-18. Look at the pedigree for Problem 11-17. Individual IV-6 said to his siblings, "Great-grandfather I-2 introduced this condition into the family -- it was never seen before his children." Is IV-6 justified in this conclusion?

Pedigree for Problem 11-14.

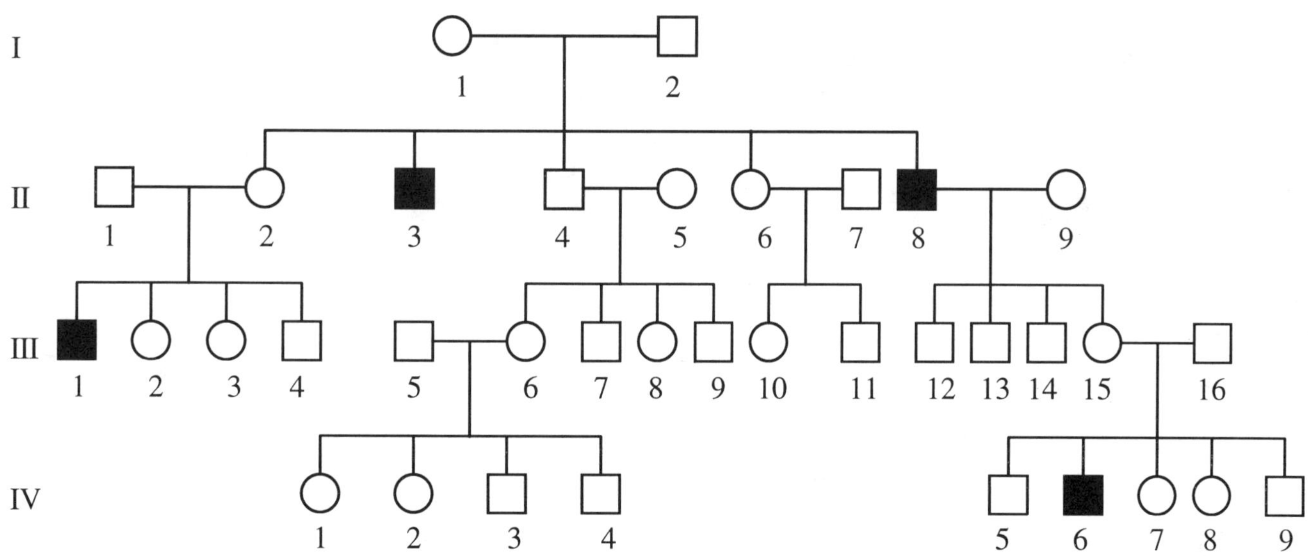

Pedigree for Problem 11-15.

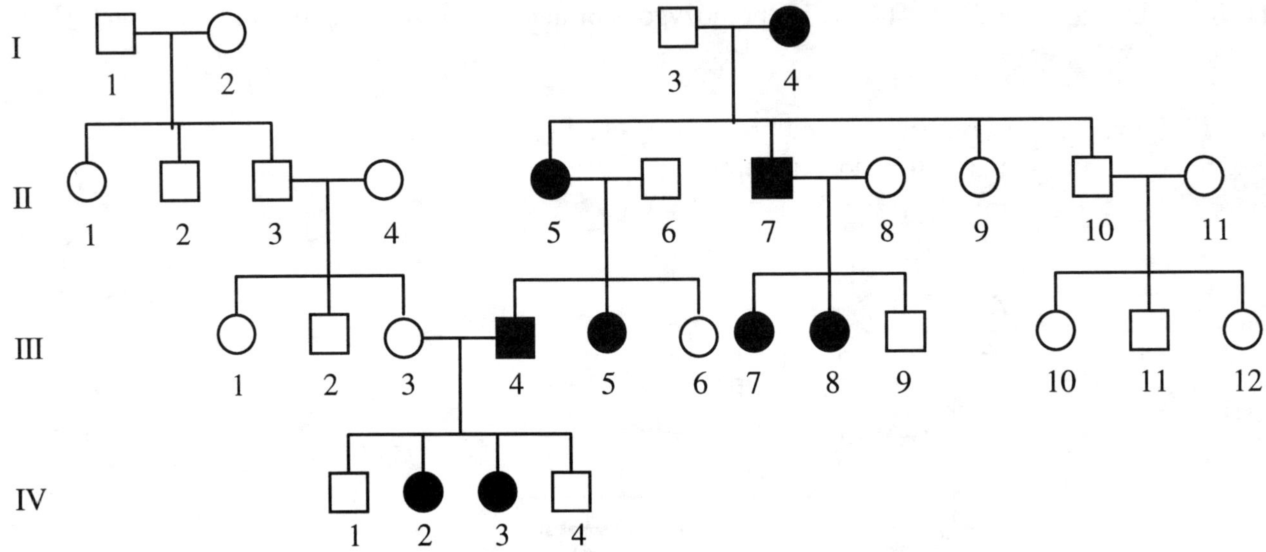

CHAPTER 6
LINKAGE

12. Some genes are linked together on the same chromosome.

So far all of the problems in this guide involve either single loci or several independently assorting loci. However, organisms have more genes than chromosomes. For example, *Drosophila melanogaster* has four pairs of chromosomes when diploid but may have 15,000 - 25,000 genes. Humans have 23 pairs of chromosomes but 60,000 - 80,000 different genes. Sooner or later geneticists encounter genes on the same chromosome. Two loci are said to be linked or to show linkage when they are on the same chromosome. Linkage brings new wrinkles to solving genetics problems, but be systematic and remember the basics and you will have no more difficulty than with any other genetics problems.

A. Use symbols that show the genes are linked.

Use a symbol system that shows that genes are on the same chromosome. If the alleles *A* and *B* are not linked, simply write the diploid genotype like this : *AA BB* or *Aa Bb*. But if they are linked, use a little line to symbolize the chromosome they have in common: *AB/AB* or *AB/ab*. Sometimes people use a double line to emphasize the diploid nature of the genotype, but that is not necessary. Because the genes are linked, the symbol *AB/ab* is different from *Ab/aB*. Both are cases of double heterozygotes, but in the first case *AB* came from one parent and *ab* from the other. In the second case, *Ab* came from one parent and *aB* from the other.

Solved Problem 12-1. Symbolize the genotype for each of the following cases.

a). A double heterozygote for the curled wing and white ocelli loci of *Drosophila* (ocelli are small light sensing organs on the top of a fly's head). Flat wing (*C*) is dominant to curled (*c*), and yellow ocelli (*W*) is dominant to white ocelli (*w*). These two loci are on the same chromosome. Assume that one parent provided the chromosome with the allele for curly wings on it and the other parent provided the allele for white ocelli.

b). The same situation as in a, but one parent provided alleles for both curled wings and white ocelli, and the other parent provided the two wild type alleles.

Answer: a). One parent provided *C* and *w*, while the other provided *c* and *W*. The symbol for this is *Cw/cW*.
b). When one parent provides both mutant alleles, the symbol is *CW/cw*.

Solved Problem 12-2. Sometimes crosses involving three linked genes are done. Yellow body (*y*), miniature wings (*m*), and forked bristles (*f*) are all linked on the X-chromosome. No alleles for these loci occur on the Y. How would you write the symbol for these three genes in a female who is heterozygous for all three and received the mutant alleles yellow and forked from one parent and the miniature allele from the other parent?

Answer: Using capital letters for the wild type alleles and lowercase letters for the mutant, this fly would be *yfM/YFm*.

PRACTICE PROBLEMS

12-3. Suppose you had a cat that was heterozygous for two loci on chromosome 3, one for the presence (*N*) or absence (*n*) of the enzyme nucleoside phosphorylase and one for presence (*H*) or absence (*h*) of the enzyme hexokinase A. If a female cat is *nH/Nh*, and her parents were both homozygotes, what were the genotypes of her parents?

12-4. Write the genotype symbols of one of the offspring of this cross:

ABc/ABc X abC/abC

12-5. Mice have fuzzy hair when the recessive "fuzzy" allele *f* is homozygous, and they display a strange stumbling sideways motion when the recessive allele tumbler (*t*) is homozygous. These genes are linked on mouse chromosome XIII. A fuzzy mouse was crossed to a tumbler, each of which was from a pure line. Use appropriate symbols to show the genotypes of the animals in this cross.

B. Detect linkage between two genes as a lack of independent assortment.

Independent assortment happens when genes are on different chromosomes. Which allele of a particular locus goes into a gamete has no influence on which allele of a second or third independently assorting locus goes into that same gamete. But if the genes are connected to one another on the same chromosome, the ocurrence of one allele in a gamete will have very much to do with the occurence of a linked allele at another locus.

For example, consider the cat in problem 12-3, with a genotype of nH/Nh. At meiosis, n is on the same chromosome as H, so they should end up in the same gametes. N and h should occur together in different gametes. So in this case, the cat could produce two kinds of gametes, (nH) and (Nh). If the genes were assorting independently, four gamete types would occur in equal numbers: (nH) (Nh) (NH) and (nh).

Solved Problem 12-6. Look at the two genotypes in Problem 12.1 a) and b). Show what kinds of gametes each would produce if the genes were completely linked together on a chromosome and what gametes they would produce if the genes were on different chromosomes and assorting independently.

Answer: The genotype in a) is *Cw*/*cW*, so the gametes produced would be *Cw* and *cW*. In b) the genotype is *CW*/*cw*. This time the gametes are *CW* and *cw*. If the genes were assorting independently, each of these genotypes would produce these four gamete types in equal proportion: *CW* *cW* *Cw* and *cw*.

Solved Problem 12-7. Look at the genotype of the fly in Problem 12-2. Show what gametes it would produce 1) if the genes were completely linked AND 2) if the three loci were independently assorting.

Answer: The genotype *YFm*/*yfM* would produce two kinds of gametes if the loci were linked, *Yfm* and *yfM*. If the genes were independently assorting, there would be $2^3 = 8$ different gametes, and they would be *YFm, yfM, YFM, yfm, yFm, YfM*., *Yfm*, and *yFM.*

Notice that in each case in which the genes are linked, the gametes that are produced are "parental" gametes. They have the same combinations of alleles as the parents did. **A clear excess of parental types among the offspring of a cross involving more than one allele is a sign of linkage.**

Practice Problems

12-8. Three alleles are suspected of being linked. A parent of genotype *AA BB CC* is crossed to a parent of genotype *aa bb cc*. (We won't symbolize them as linked until we know for sure.) What kinds of gametes can each parent produce, and what is the genotype of the progeny of this cross? Consider both the case of independent assortment and complete linkage of the genes.

12-9. What gametes would the progeny from Problem 12-8 produce if all three loci are linked?

12-10. In *Drosophila* the recessive gene for vermilion eye is inherited as we would expect an X-linked gene in *Drosophila* to be inherited. Likewise, the recessive gene for miniature wings is inherited as an X-linked gene. Are the two genes linked?

C. Genes are usually not completely linked even if on the same chromosome.

So far, in the problems you've solved involving linked genes, the two alleles stay together so that wherever one allele goes the other goes. This complete linkage happens in some cases, but it is more common to see some recombination. This happens because the two homologous chromosomes exchange material, a process called "crossing over." If the genes are linked, there will still be an excess of the parental types, but some recombinant (crossover or non-parental) types will also be seen because of crossing over. The excess of parental types is a very good indication of linkage, even if there are recombinants as well.

Solved Problem 12-11. Return to Problem 12-1 to remind yourself about the genes for curled (vs. flat) wing and white (vs. yellow) ocelli. A cross between a homozygous fly with flat wings and yellow ocelli and a fly with curled wings and white ocelli resulted in heterozygous offspring. The male heterozygotes were crossed to a females with curled wings and white ocelli to obtain another generation. Likewise, the female heterozygotes were crossed to males with curled wings and white ocelli. There is no crossing over in male *Drosophila* (which is not typical of most organisms). Diagram the P and F_1 generations of the cross involving male heterozygotes, showing the phenotypes and genotypes of each fly involved, and then show the kinds of gametes the male F_1 will produce.

Answer: P: CW/CW male X cw/cw female
flat wings — curled wings
yellow ocelli — white ocelli

F_1: all CW/cw
flat wings
yellow ocelli

Male gametes (no crossing over, so no recombinants)

1/2 (CW) 1/2 (cw)

Solved Problem 12-12. Diagram the P and F_1 generations of the cross involving female heterozygotes and show what kind of gametes the F_1 female produces. Unlike in the male, her chromosomes sometimes undergo crossing over, so that 20% of the gametes are recombinant or crossover types, and 80% are parentals.

Answer: The diagram of the P and F_1 is the same as for Problem 12-11. But the gametes produced by the F_1 are different.

Female gametes:

Parental types (80% total):
- CW 0.4 or 40%
- cw 0.4 or 40%

Recombinant types (20% total):
- Cw 0.1 or 10%
- cW 0.1 or 10%

PRACTICE PROBLEMS

12-13. Take Problems 12-11 and 12-12 one step further. Both male and female F_1 are to be crossed to cw/cw. What will be the types of offspring and the frequency of each type of offspring from the cross using the male heterozygote? The female? What would be the types and frequency of offspring from either cross if there were independent assortment?

D. Use a test cross to make it easier to detect deviations from independent assortment.

Remember that in a test cross one parent is homozygous recessive for all the genes being considered. As a result, the kinds and proportions of gametes from the other parent are the same as the kinds and proportions of offspring from the test cross. For example, if the other parent has the genotype *Aa Bb*, its gamete types will be *AB, Ab, aB* and *ab*, the *aa bb* parent will have all *ab* gametes and the offspring will be *AaBb, Aabb, aaBb* and *aa bb*. A test cross is the best way to detect deviation from independent assortment.

Solved Problem 12-14. Do a test cross of each of the following heterozygotes and show the expected types and proportions of progeny if the genes assort independently.

A. Aa Bb B. RR Ss TT Ww Zz

Answer: A . Aa Bb X aa bb

Gametes: (A B) (A b) (a B) (a b) all (a b)

Progeny Aa Bb The four types will occur in a ratio of 1:1:1:1
aa Bb
Aa bb
aa bb

B. RR Ss TT Ww Zz X rr ss tt ww zz

Gametes: RSTWZ, r STWZ; all rstwz
RSTWz, r STWz
RSTwZ, r STwZ
RSTwz, r Stwz

Progeny: Rr Ss Tt Ww Zz The eight types will occur in a ratio of 1:1:1:1:1:1:1:1.
Rr Ss Tt Ww zz
Rr Ss Tt ww Zz
Rr Ss Tt ww zz
rr Ss Tt Ww Zz
rr Ss Tt Ww zz
rr Ss Tt ww Zz
rr Ss tt ww zz

A modification of the test cross approach can be done with X-linked genes. In that case, males are hemizygous. So if you cross a heterozygous female to a male with the recessive allele on its X chromosome, the offspring will be in the same proportions as the gametes of the female. In effect, for X-linked genes, crossing a heterozygous female to a recessive male has the same usefulness as a test cross.

	Female		Male
	AB/ab	X	ab/Y
or you can write these:	$X^{AB} X^{ab}$		X^{ab} Y
Gametes:	X^{AB} (Parental)		X^{ab}
	X^{ab} (Parental)		Y
	X^{Ab} (Recombinant or crossover)		
	X^{aB} (Recombinant or crossover)		

	X^{ab}	Y
X^{AB}	$X^{AB}X^{ab}$	$X^{AB}Y$
X^{ab}	$X^{ab}X^{ab}$	$X^{ab}Y$
X^{Ab}	$X^{Ab}X^{ab}$	$X^{Ab}Y$
X^{aB}	$X^{aB}X^{ab}$	$X^{aB}Y$

See how the proportions of the different genotypes and phenotypes will be the same as the proportions of gametes from the heterozygous female.

Solved Problem 12-15. Only 3% of the gametes show recombination between vermilion (a recessive eye color allele) and miniature (a recessive wing shape mutant) in *Drosophila*. The two loci are both on the X chromosome. F_1 females from a cross of a vermilion, miniature parent and a homozygous red-eyed, normal winged parent, were crossed to vermilion, miniature males. What will be the proportions of each phenotype among the offspring? Consider males and females separately.

Female heterozygote   Male for test cross

VM/vm   X   vm/Y

Gametes: Parental types (97%)

(VM) .485   (vm) .485   (vm)   (Y)

Recombinant (crossover) types (3%)

(Vm) .015   (vM) .015

Female offspring	VM/vm	.2425	Male offspring	VM/Y	.2425
	vm/vm	.2425		vm/Y	.2425

(The 97% parentals are divided among four classes: two classes of males and two classes of females.)

Vm/vm	.0075		Vm/Y	.0075
vM/vm	.0075		Vm/Y	.0075

(The 3% recombinants are divided among four classes also.)

PRACTICE PROBLEMS

12-16. How would the answer for 12-15 differ if the original female heterozygote were *Vm*/*vM*?

12-17. You can construct the female heterozygote for problem 12-15 in two ways, starting either with *VM*/*VM* females and *vm*/Y males or with *vm*/*vm* females and *VM*/Y males. These females are identical, regardless of how they are constructed. Still, only one of the two possible crosses will produce the males that will permit the performance of a test cross of the heterozygous females. Which way works for doing the X-linked equivalent of a test cross?

CHAPTER 7
LINKAGE MAPS

13. Use the frequency of recombination or crossing over between genes to measure how far apart the loci are on their chromosome.

The percentage of recombinant types is a characteristic of the two linked genes being studied. As you will see in the following examples, each pair of genes we look at will have its own unique frequency of crossover types. Since the genes are arranged linearly on a chromosome, we can use the frequency of crossing over to make a "linkage map" of linked genes. Geneticists assume that crossing over can occur at any point along a chromosome, so the farther apart two genes are, the more likely it is that a crossover between them will produce recombinant gametes.

A. The unit of distance between linked genes is the percent crossing over.

A road map shows the distance between two points in miles or kilometers; treasure maps show distances in paces. Linkage maps use % crossing over as their unit of distance. On a linkage map, **one map unit** is **1% crossing over**. You can find the % crossing over by doing a test cross of a heterozygote for two linked genes. The % crossing over is (the number of crossover types) ÷ (the total number of offspring) X 100.

Solved Problem 13-1. Many genes are known on the X chromosome of *Drosophila*, including cut wings (*ct*), vermilion eyes (*v*) and miniature wings. In the appropriate cross, crossovers occur 3.1% of the time between *m* and *v* and 13% of the time between *ct* and *v*. Which wing gene is closest to gene *v*?

Answer: The distance between miniature and vermilion is much smaller (3.1%) than the distance between cut and vermilion (13%). So *m* is closer to *v* than *ct* is.

Solved Problem 13-2. A recessive allele (*a*) in corn increases resistance to grasshoppers, and another recessive allele (*z*) causes stripes on the leaves. A homozygous *aa* plant was crossed to a homozygous *zz* plant, and then the double heterozygote was test-crossed. Of 5280 corn plants resulting from this cross, 2480 were *aZ*/*az*, 2536 were *Az*/*az*, 140 were *AZ*/*az*, and 124 were *az*/*az*. What is the map distance between these two genes?

Answer: The *AZ*/*az* and *az*/*az* represent the crossover types. Of 5280 plants, 264 are recombinant. 264/5280 = .05 and .05 X 100 = 5%. The map distance is 5 map units.

PRACTICE PROBLEMS

13-3. White *(w)* is a recessive eye color allele on the X chromosome of *Drosophila*, and miniature (*m*) is a recessive wing shape allele also on the X chromosome. So we know they're linked. When a cross of *wm*/*WM* X *wm*/Y was carried out, 1290 progeny were obtained, of which 843 were either red, normal or white, miniature and 447 were either miniature with red eyes or white with normal wings. What is the map distance between white and miniature?

13-4. If the cross *wM*/*Wm* X *wm*/Y were done instead of the cross described in 13-2., what proportion of the progeny would be white, miniature or red, normal? Why?

B. A map of a chromosome can be made by doing a set of pair-wise crosses.

If we have three linked genes, we can make a map that places all three in their relative order along the chromosome and that shows the map distances (% crossover) between the genes. An orderly strategy for building a gene map from data on crosses of pairs of genes will help you solve most problems.

1). Determine the % crossing over between all three pairs of genes involved.

2). The pair with the longest map distance must be on the outside.

3). The two shorter distance are the distances from the outside genes to the middle one.

4). Check the order to see if the total distance between each outside gene and the middle gene is approximately equal to the distance between the outside genes.

Solved Problem 13-5. In a certain organism, crossing over between *R* and *S* occurs 15% of the time, and crossing over between *S* and *T* occurs 22% of the time. Between *R* and *T* crossing over happens 7% of the time. What is the order and map distances for a map of these three genes?

Answer: In this case we know the genes are linked since they all show a small percentage of crossover types. So step 1 is done. Step 2: The largest map distance is 22 units between *S* and *T*, so those are the outside genes. *R* must be in the middle, (Step 3) 15 units from *S* and 7 units from *T*. Here is the map:

S------15---------R---7----T.

Step 4: The sum of *SR* and *RT* is 22, which is what was observed, so this probably is the map.

Solved Problem 13-6. Pair-wise crosses involving three different *Drosophila* mutations on chromosome II produced three heterozygotes, which were then test crossed.

The cross involving black and vestigial resulted in 20,153 flies, of which 3,578 were recombinant. The cross involving black and purple produced 48,931 progeny of which 3,026 were recombinant. Finally, the cross involving vestigial and purple produced 2,598 recombinants out of 13,601 flies. Map these genes.

Answer: Step 1: The % crossing over in each case is black - vestigial 3578 ÷ 20153 X 100 = 17.8%. Between black and purple 3026 ÷ 48931 X 100 = 6.2%. Between vestigial and purple 2598 ÷ 13601 X 100 = 19.1%
Step 2: Vestigial and purple are farthest apart at 19.1 map units, so they are the outside genes.
Step 3: The distances to the outside genes from the middle give this map:

vestigial -----17.8---- black ----6.2---- purple.

Step 4: The sum of the short distances is 17.8 + 6.2 = 24, slightly more than 19.1, but it's probably right.

PRACTICE PROBLEMS

13-7. For these three pairs of linked genes, construct a linkage map showing the order and distance between the loci.

r -- s 10 % recombinants.
s -- t 20% recombinants
r -- t 15% recombinants

13-8. In *Geneticus impossibili*, an organism found only in genetics problems, the antennae may be long or short; the eye facets may be gold or silver; and the rearmost bristles may be set close together or far apart. A homozygous long, gold, far-set individual was mated to a homozygous short, silver, close-set one. The resulting hybrids were testcrossed. The data were then divided up into pair-wise categories like this:

long, gold	296	long, far	444	gold, far	270
long, silver	198	long, close	50	gold, close	228
short, gold	202	short, far	50	silver, far	224
short, silver	304	short, close	456	silver, close	278

Construct a gene map.

C. An easier way to make a map is to do a test cross involving three linked genes.

Constructing a map of three genes from pair-wise crosses is not as economical as doing a single cross involving all three genes. Usually the cross is a test cross of the appropriate heterozygote for the linked genes and is called a three point test cross. Just as with other problems you've encountered in this book, you can analyze such crosses by breaking them down into simpler problems -- three problems using pairs of linked genes. There are methods for analyzing all three genes at once, but you can't go wrong if you treat these as three separate problems and then combine them.

1). **Decide if all three loci are indeed linked.** Do this by looking at each possible pair of loci to see if there is an excess of some of the classes of offspring instead of independent assortment.

Solved Problem 13-9. Corn seeds may be colored (*C*) or white (*c*); full (*S*) or shrunken (*s*), and starchy (*W*) or waxy (*w*). A plant heterozygous for all three loci was produced by crossing two homozygotes, and the resulting heterozygote was test crossed . Here are the kinds and numbers of offspring:

White, shrunken, starchy	116
White, shrunken, waxy	2
White, full starchy	626
White, full, waxy	2708
Colored, shrunken, starchy	2538
Colored, shrunken, waxy	601
Colored, full, starchy	4
Colored, full, waxy	113

Are the three loci linked?

Answer: There are 3139 colored, shrunken and 3334 white, full but only 118 white, shrunken and 117 colored, full. That is a definite excess of parental types, so these loci are linked.

There are 2654 shrunken, starchy, 2821 full, waxy, 630 full, starchy, and 603 shrunken, waxy, so again there is an excess of parentals and the loci are linked.

Finally 2542 colored, starchy and 2821 white, waxy are clearly in excess compared to 714 colored waxy and 742 white starchy. So this pair of loci also is linked. All three loci are on the same chromosome.

Solved Problem 13-10. Female Drosophila heterozygous for three X-linked loci were crossed to males having recessive alleles at all three loci. The following listing shows the male and female offspring separately; the male progeny have a Y chromosome; female progeny have two alleles for each locus.:

RST/rst	1243	rST/rst	5
R S T/Y	1187	r S T/Y	8
RSt/rst	926	rSt/rst	9111
R S t/ Y	889	r S t/ Y	8901
RsT/rst	8820	rsT/rst	942
R s T/Y	9215	r s T/Y	907
Rst/rst	8	rst/srt	1234
R s t/Y	7	r s t / Y	1306

Show that these genes are all linked.

Answer: Each class is made up of both males and females, so add them together in the tabulation. For example, both *Rst*/*rst* and *R s t*/Y received an *R s t* chromosome from the heterozygous parent in the test cross, so they should be added together in the following computations.

For *R* and *S*, two classes (*Rs* and *rS*) are very much in excess.:

R S =1243 + 1187 + 889 + 926 = 4245
Rs = 8820 + 9215 + 8 + 7 = 18,050
rS = 5 + 8 + 9111 + 8901 = 18,025
rs = 942 + 907 + 1234 + 1306 = 4389
So these loci are linked.

For *R* and *T*, RT = 20,465
Rt = 1830
rT = 1862
rt = 20,552. These loci are linked since two classes are in excess.

For *S* and *T*, ST = 2443
St = 19,827
sT = 19,884
st = 2555
Again two classes are in excess, so the genes are probably linked.

2). Figure out the map distance between each pair of genes.

From here on these problems are just like those in the previous section. You've already grouped the pairs and identified which are parental and which recombinant classes. So now just calculate the percentage of recombinants and you'll have what you need to construct the map.

Solved Problem 13-11. For Problem 13-9, construct a map based on the data in the answer to that problem.

Answer: The distance between the white/color locus and the shrunken/full locus is (235 ÷ 6708) X 100 = 3.5% or 3.5 map units.

The distance between the shrunken/full locus and the strachy/waxy locus is (1233 ÷ 6708) X 100 = 18.4% or 18.4 map units.

The distance between the white/color locus and the starchy/waxy locus is (1456 ÷ 6708) X 100 = 21.7% or 21.7 map units.

Since 21.7 is the largest map distance, white/color and starchy/waxy are the outside genes, and the map is:

white/color----3.5----shrunken/full---------------18.4--------------starchy/waxy

Solved Problem 13-12. Use the data in the answer to Problem 13-10. to construct a map of the *R, T* and *S* loci of *Drosophila.*

Answer: The total number of plants among the progeny is 44,709. For the *R* - *S* distance, there are 8634 recombinants (*RS* and *rs*). So the map distance is (8634 ÷ 44709) X 100 = 18.71 map units.

For the *R* -*T* distance, (3692 ÷ 44709) X 100 = 8.3 map units

For the *S* - *T* distance, (4998 ÷ 44709) X 100 = 11.2 map units.

The outside loci must be *R* and *S* with the largest distance between them.. So the map is

R ------8.3----- T ----------11.2 ---------- S

PRACTICE PROBLEMS

13-13. A three point test cross was done using a plant heterozygous for the genes M, N and P. One thousand offspring were obtained. The numbers of each phenotype are shown in the following table.

MNP	343	mNP	9
MNp	66	mNp	61
MnP	79	mnP	74
Mnp	11	mnp	357

a) Write the symbol for the genotype of the heterozygous parent.

b) If the heterozygous parent was obtained from a cross of two homozygous plants, what were their genotypes?

c) Construct a map of these three genes showing their order and the distance between them.

13-14. Heterozygotes for three loci on the second chromosome in flies were involved in a three-point test cross. From the data on progeny that resulted from this cross, determine the linkage map for these three genes.

DPE/dpe	1951	DpE/dpe	40
dpe/dpe	1910	dPe/dpe	49
DPe/dpe	179	Dpe/dpe	470
dpE/dpe	180	dPE/dpe	501

ANSWERS TO PRACTICE PROBLEMS

CHAPTER 1 SEVEN SYSTEMATIC STEPS

1-2. P: brown X white

F_1 all brown

F_2 1 white, the others brown

1-3. P: yellow, plain X white, marked

F_1	yellow, marked	yellow, plain	white, marked	white, plain
	65	56	61	59

1-4. P: golden eyed X green eyed

F_1: some golden-eyed; some green-eyed

2-3. If the Huntington's locus is *H* for disease and *h* for normal and cystic fibrosis is *C* for normal and *c* for cystic fibrosis, then

A) *Hh* B) *hh CC* or *hh Cc* C) *hh cc*

2-4. P: EE X ee

F_1 all Ee

2-5. The letter G seems natural for this cross. Let *G* be the dominant allele and *g* be the recessive. The golden-eyed lacewings are *Gg*, and the green eyed lacewings are *gg*.

3-2. Black sheep and spotted goats were true breeding, while solid colored goats and white sheep were not.

3-3. The parents are not homozygous since they gave birth to pups unlike themselves. The pups, however, are true breeding. They always produce silver-backed pups when they are mated with each other. So the parents are heterozygous and the pups homozygous.

3-6. Yes, we can know the genotypes of the true breeding strains since "true breeding" means "homozygous." If we let the symbol *B* stand for the allele for white sheep and *b* stand for the allele for black sheep, then Jake's black sheep were *bb* and the white sheep were *B* _ (where the blank means that the other allele could be either *B* or *b*). Likewise, if *S* is the allele for solid colors, and *s* is the allele for spotted, Jake's spotted goats are *ss*, and the solid goats are *S* _.

3-7. If peroneal muscular atrophy were due to a recessive allele, it would be possible for two parents without the disease to give birth to an affected child if both of them were heterozygous. If at least one parent has to have the disease for the child to inherit it, then the allele must be dominant.

3-10. Since the flies have the dominant phenotype, they must either be *WW* or *Ww*. We can know for certain that they have at least one *W* allele. So we could symbolize their genotype as *W* _, with the blank standing for either *W* or *w*.

3-11. The runt litter mates are *rr*, which means that each parent was *Rr*. Since Honey is not a runt, he got an *R* allele from one parent, but he could have received either an *R* or an *r* from the other parent. So he is either *RR* or *Rr*.

3-12. Cross Honey to a runt pig (*rr*). If Honey is *RR*, all the offspring will get an *R* from Honey and will therefore all be normal. But if Honey is *Rr*, some offspring will get an *r* from Honey. They must get an *r* from the runt, so some of Honey's offspring will be *rr* runts.

3-13. Genotypes a, d, and e are homozygous for at least one of the recessive alleles, so they are each sensitive to potassium. Genotypes b and c have at least one dominant allele at each locus, so they are resistant.

4-2. All are true breeding (homozygous) so if their genotype is *WW*, all the gametes will be *W*. If their genotype is *ww*, all their gametes will be *w*.

4-3. Both parents are heterozygous, so each will produce *A* and *a* gametes in equal proportions.

4-4 If he uses a male parent, the allele will be lost. The gametes are haploid. If that allele is lethal, as in this case, it will kill the sperm. No sperm with that lethal allele will be able to fertilize any eggs. A heterozygous female could produce eggs with the gene which would then be passed on to the next generation. But it could only be maintained in heterozygous females because male parents will never pass the gene on. So if there is a way to identify the females with the allele, the gene could be maintained.

4-8. Both parents must be heterozygous for cleft palate. Let *P* be normal and *p* be the cleft palate allele. Each pig can produce both *P* and *p* gametes in a 1:1 ratio.

4-9.

EFTV	eFTV
EFTv	eFTv
EFtV	eFtV
EFtv	eFtv
EfTV	efTV
EfTv	efTv
EftV	eftV
Eftv	eftv

4-10. If there are 10 heterozygous loci, there would be 2^{10} or 1024 possible gametes. Aren't you glad I didn't ask you to figure out what they all are!

5-3. Each individual only produces one kind of gamete, so the Punnett square is simple:

	M N
m n	

5-4. *Bb Ff Gg* individuals will produce 8 kinds of gametes. So there will be 8 gametes on each side of the square.

	BFG	BFg	BfG	Bfg	bFG	bFg	bfG	bfg
BFG								
BFg								
BfG								
Bfg								
bFG								
bFg								
bfG								
bfg								

5-5. The one parent with 14 heterozygous loci would produce 2^{14} or 16,384 gametes and the other parent with 3 loci heterozygous would produce 8 kinds of gamete. Since a

Punnett Square has gametes along each side, this one would be 16,384 squares in one direction and 8 squares in the other.

6-2.

	M N
m n	Mm Nn

6-3. This one will have 8 X 8 = 64 intersections!

	BFG	BFg	BfG	Bfg	bFG	bFg	bfG	bfg
BFG	BB FF GG	BB FF Gg	BB Ff GG	BB Ff Gg	Bb FF GG	Bb FF Gg	Bb Ff GG	Bb Ff Gg
BFg	BB FF Gg	BB FF gg	BB Ff Gg	BB Ff gg	Bb FF Gg	Bb FF gg	Bb Ff Gg	Bb Ff gg
BfG	BB Ff GG	BB Ff Gg	BB ff GG	BB ff Gg	Bb Ff GG	Bb Ff Gg	Bb ff GG	Bb ff Gg
Bfg	BB Ff Gg	BB Ff gg	BB ff Gg	BB ff gg	Bb Ff Gg	Bb Ff gg	Bb ff Gg	Bb ff gg
bFG	Bb FF GG	Bb FF Gg	Bb Ff GG	Bb Ff Gg	bb FF GG	bb FF Gg	bb Ff GG	bb Ff Gg
bFg	Bb FF Gg	Bb FF gg	Bb Ff Gg	Bb Ff gg	bb FF Gg	bb FF gg	bb Ff Gg	bb Ff gg
bfG	Bb Ff GG	Bb Ff Gg	Bb ff GG	Bb ff Gg	bb Ff GG	bb Ff Gg	bb ff GG	bb ff Gg
bfg	Bb Ff Gg	Bb Ff gg	Bb ff Gg	Bb ff gg	bb Ff Gg	bb Ff gg	bb ff Gg	bb ff gg

6-4. The square is 16,384 X 8. So the number of intersections is 16,384 X 8 = 131,072.

7-3. For any one allele, a cross of two heterozygotes will produce homozygous recessives 1/4 of the time. So the chance of being homozygous for recessive alleles at 10 different loci is 1/4 X 1/4 X 1/4 X 1/4 X 1/4 X 1/4 X 1/4 X 1/4 X 1/4 X 1/4 or $(1/4)^{10}$, which is 1/1,048,576. (Think if you had to do this with a Punnett square!)

7-4. Diagram the cross, including all the information in the problem:

P:	Rr hh	X	rr Hh
	Black and white		Red and white
	Horned cow		Hornless bull

Gametes: 1/2 (Rh) 1/2 (rh)   1/2 (rH) 1/2 (rh)

Punnett Square

	Rh	rh
rH	Rr Hh black/white hornless	rr Hh red/white hornless
rh	Rr hh black/white horned	rr hh red/white horned

Red and white hornless occur 1/4 of the time.

There are 2 X 15 = 30 offspring. So 1/4 of 30 or about 6 or 8 offspring will be red and white, hornless.

The father is red and white hornless. Red and white hornless also occurs 1/4 of the time, so about 6 or 8 of the 30 progeny will be like that.

7-5 The proportion of offspring which are *Aa* will be 1/2, since the parents were both heterozygous for that locus. In a similar way, the proportion that will be *Dd* and *Ee* will also be 1/2. Since *BB* is homozygous in both parents, all offspring will be *BB*. Likewise all the offspring will be *cc*, *ff*, *gg* and *HH*. Thus the proportion that will be like the parents will be 1/2 X 1/2 X 1/2 X 1 X 1 X 1 X 1 X 1 = 1/8.

CHAPTER 2 GOING BEYOND THE SEVEN STEPS

8-3. In the cross of *Ff Gg Hh* X *ff gg hh*, the triple heterozygote will produce 8 different gamete types, and the triple homozygote will produce only one kind (*f g h*). So each of the possible offspring will receive one of the eight possible gametes form the heterozygote and one *f g h* gamete. All of the possible offspring are listed below.

Ff Gg Hh	ff Gg Hh
Ff Gg hh	ff Gg hh
Ff gg Hh	ff gg Hh
Ff gg hh	ff gg hh

8-4. Let's let *D* be the allele for polydactyly and *d* its recessive while *p* is the allele for phenylketonuria and *P* is the allele for lack of phenylketonuria. Since the man's father had phenylketonuria, the father must have been *pp*, so the man is *Pp*. The man doesn't have the dominant polydactyly either, so he is *dd*. The woman has polydactyly, so she is *Dd*. Since the problem specifies that she has no phenylketonuria allele, she is also *PP*. Here is a diagram of what we know about this cross:

P: Pp dd X PP Dd

Gametes: 1/2 (Pd) 1/2 (pd) 1/2 (PD) 1/2 (Pd)

Punnett square:

	Pd	pd
PD	PP Dd polydactyly	PpDd polydactyly
Pd	PP dd Neither condition	Pp dd Neither condition

There is no chance that any offspring will have both conditions, nor is there a chance that any offspring will have phenylketonuria. Half the offspring can be expected to have polydactyly.

8-5. The investigators have the two single mutant lines *LL pp* and *ll PP*. Cross each of them to every unknown. Each type of unknown will give different results. Diagram the crosses to convince yourselves of the results summarized in the table. One cross is diagrammed to give you an example of what to do.

Example:

P: ll PP (known) X unknown (if ll PP)

Gametes all (lP) all (lP)

F_1: all ll PP light eared

P: ll PP (known) X unknown (if LL pp)

Gametes all (lP) all (Lp)

F_1: all Ll Pp normally eared

Known testers	Unknowns		
	if ll PP	if LL pp	if ll pp
ll PP	all light ears	no light ears	all light ears
LL pp	no light ears	all light ears	all light ears

So the double mutants are the unknowns that have light eared progeny in both crosses to the testers. The *ll* single mutant is the one that produces light eared progeny when crossed to known *ll* testers, and the *pp* single mutant is the one that produces light eared progeny when crossed to the known *pp* tester.

8-8. The ratio is nearly 9:3:3:1, which is characteristic of a cross of two parents heterozygous for two independently assorting genes. So the parents were

Dd Nn	X	Dd Nn

8-9. Write what we know of the genotypes of Izzy and Lucinda.

Izzy	Lucinda
B_ ww F_	bb W_ ff

If Lucinda's other parent had big white spots, weak hooves and at least one *f* allele (B_ W_ _f), then it is possible that Izzy could be Lucinda's parent.

8-10. Look for familiar ratios. The F_2 progeny approximate a 9:3:3:1 ratio, characteristic of a cross of two double heterozygotes with both loci showing dominance. So let the F_1 be Aa Bb. Crossing two of those gives the F_2.

311	A_ B_	(normal green)
107	aa B_	(lighter green)
123	A_ bb	(pale green)
32	aa bb	(yellow green)

Then Strain T must be *aa BB*, given the genotype of the lighter green F_2. Strain S must be *AA bb*.

CHAPTER 3 MORE COMPLEX SITUATIONS

9-3. A. *Ss* is intermediate in phenotype between *SS* and *ss*, so this is a case of incomplete dominance.

B. A 3:1 ratio means the black homozygote and the heterozygote have the same phenotype. This is a case of dominance.

C. *D'D* has an effect midway between *DD* and *D'D'*. So incomplete dominance.

D. The heterozygote has both phenotypes, M and N, so this is a case of codominance.

E. With regard to the lethal effect, the heterozygote and the wild type homozygote are the same. Thus this is dominance. With respect to the color effect, there is no telling. since there is no color of the Montezuma homozygote for comparison.

9-4. Results will be the same. The F_1 radishes will still be heterozygous for both loci.

9-5. We can use the Product Rule on this one. Consider the cross as two separate crosses, C^RC^W X C^RC^W and Hh X Hh.

The proportions that have horns and are roan is 1/4 (those with horns) X 1/2 (those that are roan). 1/4 X 1/2 = 1/8.

The proportion that lack horns and are white is 3/4 X 1/4 = 3/16.

9-9. For an i^oi^o child (type O) to appear among the offspring, each parent must contribute an i^o allele. For AB to appear among the offspring, one parent must contribute an I^A and one must contribute an I^B. One parent must be I^Ai^o and one must be I^Bi^o. The cross is thus the same as in Problem 9-7, and type O offspring will occur 1/4 of the time.

9-10. P: $C\,c^{ch}$ X $c^h\,c$

Normal Himalayan

Gametes: 1/2 (C) 1/2 (c^{ch}) 1/2 (c^h) 1/2 (c)

Punnett Square:

	C	c^{ch}
c	Cc normal	$c^{ch}c$ chinchilla
c^h	$C\,c^h$ normal	$c^h\,c^{ch}$ Himalayan

Such a cross would result in half normal progeny, 1/4 chinchilla. and 1/4 Himalayan

9-11. A. Holstein must be *ss* since *s* is recessive to all the other alleles. So two *ss* cattle could only produce *ss* offspring. Holsteins can only produce Holsteins.

B. Two Herefords could each be s^hs. Then:

$s^h s$ X $s^h s$

	s^h	s
s^h	$s^h s^h$ Hereford	$s^h s$ Hereford
s	$s^h s$ Hereford	$s\ s$ Holstein

So Herefords could produce Holsteins.

C. Dutch belts will be *S_*, where the blank could be any of the four alleles. Holsteins are *ss*, so for Dutch belts to have Holstein offspring, both parents would have to be *Ss*. Solid color offspring could either be $s^c s^c$ or $s^c s$. At least one parent would have to be Ss^c. So both solid color and Holstein could not result from mating the same parents. These are multiple alleles of *one* locus, and no parent can have more than two alleles.

9-14. When we cross two double heterozygotes, *Aa Bb*, for example, we get this ratio:

9 A_ B_
3 aa B_
3 A_ bb
1 aa bb

When we do a test cross of *Aa Bb* to *aa bb*, the *Aa Bb* parent produces four gamete types in equal proportion: *A B*, *a B*, *A b*, and *a b*. So four classes of offspring occur in equal proportion:

1 Aa Bb
1 aa Bb
1 Aa bb
1 aa bb

The modifications come from gene interactions that result in more than one of the classes having the same phenotype.

A. A 12:3:1 ratio is obtained from the 9:3:3:1 in this way:

9 A_ B_ | These two genotypes have the same phenotype.
3 A_ B_ | Allele *A* is epistatic to the B-locus.
3 aa bb
1 aa bb

So the modified 1:1:1:1 ratio is

1 Aa Bb	These two genotypes would have the same phenotype,
1 Aa Bb	so the ratio is 2:1:1.
1 aa bb	
1 aa bb	

B. A 9:7 ratio is obtained in this way:

9 A_ B_	
3 aa B_	These three genotypes would have the same phenotype
3 A_ bb	if a dominant allele were required at both loci to produce
1 aa bb	one of the phenotypes.

So the modified 1:1:1:1 ratio is:

1 Aa Bb	
1 aa Bb	These three genotypes have the same phenotype, so the
1 Aa bb	modified ratio is 1:3
1 aa bb	

C) A 15:1 ratio is obtained in this way:

9 A_ B_	These three classes would have the same phenotype if one
3 aa B_	of the phenotypes required that both loci be homosygous
3 A_ bb	recessive.
1 aa bb	

So the modified 1:1:1:1 ratio is:

1 Aa Bb	These three classes have the same phenotype, so the
1 aa Bb	modified ratio is 3:1.
1 Aa bb	
1 aa bb	

9-15. The problem involves a gene that is not quite completely epistatic. Allele *b* suppresses, but does not eliminate, the stripe produced by allele *S*.

F_1	All Bb Ss	tan with a stripe	
F_2	B_ S_	tan with a stripe	9/16
	B_ ss	tan with no stripe or spot	3/16
	bb S_	brown with a spot	3/16
	bb ss	brown with no stripe or spot	1/16

Since epistasis is not complete, it's the same old double heterozygote problem with a 9:3:3:1 ratio. This emphasizes that the segregation and assortment of genes is not changed by epistasis. The only new issue is remembering how the genes interact to produce phenotypes.

9-16. Diagram what you know from how the problem is stated.

P:	black X hairless
F_1	All black
F_2	Black 100 Hairless 43 Brown 30

The F_2 ratio is close to a 9:4:3 which is a modified 9:3:3:1. So the F_1 must be double heterozygotes. Let *B* be black and *b* brown. Let *H* be haired and *h* be hairless. Then the F_1 and F_2 are:

F_1	All black	Bb Hh
F_2	Black 100 Hairless 43 Brown 30	B_ H_ 9/16 B_ hh 3/16 and bb hh 1/16 bb H_ 3/16

The parents must have been Black *BB HH* and Hairless *bb hh*.

9-17. As always, a cross of two double heterozygotes will result in a ratio of

9 A_ B_
3 aa B_
3 A_ bb
1 aa bb

The problem indicates that at least one dominant allele must be present at each locus to get black spines, so only the *A_ B_* class will have black spines. This becomes a 9 black: 7 white ratio. Of 1000 plants, 9/16 or 563 will have black spines and 7/16 or 438 will have white spines.

CHAPTER 4 SEX LINKAGE

10-4. If the mother is heterozygous, she will have normal color vision. Half of her offspring will receive her X-chromosome which includes the color-blindness gene. Each male that receives that X will receive a Y chromosome from the father and so will be color blind. Each female that receives the color blindness gene from the mother will receive a normal X from the father, so they will not be color blind.

10-5.

	Male		Female
	Vv X^fY	X	Vv X^FX^f
	Male		Female
	Normal wings		Normal wings
	Forked bristles		Normal bristles
Gametes:	1/4 VX^f		1/4 VX^f
	1/4 vX^f		1/4 vX^f
	1/4 VY		1/4 VX^F
	1/4 vY		1/4 vX^F

	V X^f	v X^f	VY	vY
V X^f	VV $X^f X^f$ Female Normal wings Forked	Vv $X^f X^f$ Female Normal wings Forked	VV X^fY Male Normal wings Forked	Vv X^fY Male Normal wings Forked
v X^f	Vv $X^f X^f$ Female Normal wings Forked	vv $X^f X^f$ Female Vestigial Forked	Vv X^fY Male Normal wings Forked	vv X^fY Male Vestigial Forked
V X^F	VV $X^F X^f$ Female Normal wings Normal	Vv $X^F X^f$ Female Normal wings Normal	VV X^FY Male Normal wings Normal	Vv X^FY Male Normal wings Normal
v X^F	Vv $X^F X^f$ Female Normal wings Normal	vv $X^F X^f$ Female Vestigial Normal	Vv X^FY Male Normal wings Normal	vv X^FY Male Vestigial Normal

Among the males, 6 have normal wings and 2 have vestigial wings for a 3:1 ratio. Also, 4 have normal bristles and 4 have forked bristles for a ratio of 1:1.

Among the females, there is also a 3:1 ratio of normal winged to vestigial and a 1:1 ratio of forked to normal bristles.

Of 400 flies, 200 are expected to be females, and of those 3/8 or 75 will have both normal wings and normal bristles, 3/8 or 75 will have normal wings and forked bristles, 1/8 or 25 will have vestigial wings and normal bristles and 1/8 or 25 will have vestigial wings and forked bristles.

Similarly the males are expected to include 75 normal, normal; 75 normal, forked; 25 vestigial, normal; and 25 vestigial, forked.

10-6. If the gene is x-linked, then Alexis must have been X^hY, and he must have received that X^h from his mother. She and Alice must both have been heterozygotes since neither had hemophilia. The most probable thing is that Alexandra and Alice inherited the gene from a common ancestor, so it is more probable that they are the cousins.

10-9. The difference in reciprocal crosses suggests X-linkage. Let *B* symbolize the Bar allele and *b* symbolize the normal. Put the alleles on the X-chromosome and the results are explained.

P:	$X^b X^b$ Female Normal	X	X^BY Male Bar	$X^B X^B$ Female Bar	X	X^bY Male Normal
F_1	Males all X^bY (normal) Females $X^B X^b$ (bar)			Males X^BY (bar) Females $X^B X^b$ (bar)		

10-10. To decide if body color is X-linked, just look at the results for body color. The cross reduces to:

P:	black-bodied male X yellow-bodied female	
F_1:	all yellow-bodied (yellow must be dominant to black)	
F_2:	males: 3 yellow: 1 black	females: 3 yellow: 1 black

Compare the results with and without sex linkage:

	If Autosomal	If X-linked
P:	bb X BB	X^bY X X^BX^B
F_1:	all Bb (yellow)	Males: X^BY (yellow) Females: X^BX^b (yellow)
F_2:	1 BB: (yellow) 2 Bb: (yellow) 1 bb (yellow)	Males: 1 X^BY (yellow): 1 X^bY (black) Females: 1 X^BX^B: 1 X^BX^b (all yellow)

If the body color gene were X-linked, we would see black bodies only among the F_2 males. But we see a 3:1 ratio with each sex. So the body-color gene is not on the X chromosome.

10-11. Look at the characters one at a time.

1. Long crossed with short gives a ratio of 1:1 in a 1:1 sex ratio. We can therefore assume that this locus is not X-linked and that one parent is a heterozygote (*Ll*) and one is a homozygote recessive (*ll*).

2. Active crossed with sluggish results in all active progeny regardless of sex. It is a result of a mating (*AA* X *aa*) where active (*A*) is dominant to sluggish (*s*).

3. Purple males mated with yellow females results in yellow males and purple females. Color segregates by sex, so we can deduce that the color locus is X-linked. Purple (X^P) is X-linked dominant over yellow (X^p). $X^p X^p$ X X^PY

CHAPTER 5 PEDIGREES

11-3. Here is the pedigree:

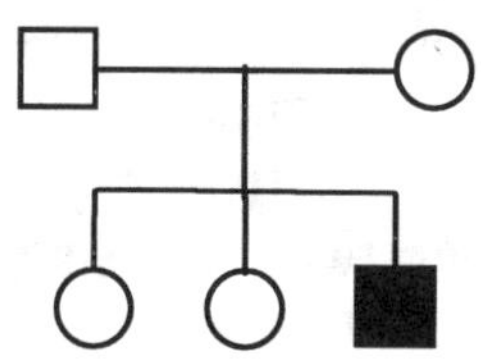

11-4. Assuming "attached ear lobes" is the character under study, shade the symbols of those with attached ear lobes. The pedigree:

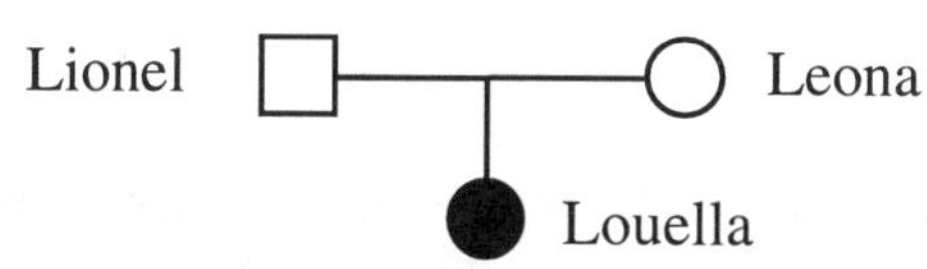

11-5. Here is the pedigree with polydactyly shaded in:

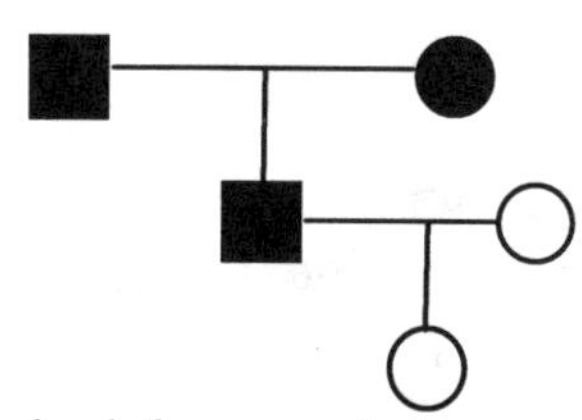

11-9. Free ear lobes is dominant. Both Lionel and Leona had free ear lobes yet gave birth to a child with attached ear lobes. This is a sign that attached is recessive.

11-10. The man must be heterozygous. The daughter by the second marriage has pituitary dwarfism, a recessive character. She must have received a recessive allele from each parent. We know she got one from her homozygous mother, but she also must have a recessive allele from the father. Since he is normal, he must also have a dominant allele, and thus is heterozygous.

11-11. The data do not allow a decision on whether polydactyly is dominant or recessive. The first mating in which two affected individuals have an affected child could happen if the parents were both homozygous recessives or if they each had at least one dominant allele.

If the condition were dominant, one or both parents could be heterozygous dominant, and they could give rise to a heterozygous child. Then the child in the third generation could have received a recessive allele from each parent.

Or, if the condition were recessive, the unaffected mother would have given a dominant allele (no polydactyly) to her daughter.

If the couple in the second generation had a daughter with polydactyly, that would not clear up the ambiguity: If the condition is dominant, and the affected father in the second generation were heterozygous, he could give the normal allele to the first daughter and the polydactyly allele to the second. If the condition is recessive, the mother in the second generation could be heterozygous and give the polydactyly allele to one child and the normal (dominant) allele to the other.

More information is needed beyond this small pedigree.

11-12. The disability is probably due to a dominant allele. Once it appears in the family, it shows up in every generation (II - V). Affected parents have both unaffected and affected children (II-1, III-2, III-3), but two unaffected parents never have affected children (I-1 and I-2; II-6 and II-7; III-9 and III-10).

11-13. If the disability is due to a dominant allele, parents with the disability will have unaffected children only if the parent is heterozygous, because the parent must give a recessive allele to unaffected children. Both II-1 and III-3 have at least one unaffected child and so are heterozygous. If a person with the dominant allele has a normal parent, then that parent must have given a recessive allele to all offspring, and the affected children will be heterozygous. III-3, III-5, IV-9 and V-2 all have a normal parent, so all must be heterozygous.

11-16. Heterozygous. The phenotype of interest is due to a recessive gene. It shows up in males of the next generation, so the allele comes from the female. She has the normal phenotype, however, so she must have a dominant allele also.

11-17. Recessive and X-linked. II-5 and IV-2 arise from two unaffected parents. This is a good indication that the gene is recessive. All the affected individuals are males. The affected individuals in generation IV all have mothers with an affected father. These are good signs of sex-linkage.

11-18 Since the allele is recessive and sex-linked, a male can not receive the gene at issue from his father, since he gets a Y chromosome from the father. II-5 had to receive his allele from his mother. IV-9 is wrong about great grandfather.

CHAPTER 6 LINKAGE

12-3. The cat must have received an *nK* from one parent and an *Nk* from the other parent. Since the parents are both homozygotes, they must be *nK/nK* and *Nk/Nk*.

12-4 The offspring will receive one chromosome from each parent, so the genotype is *ABc/abC*.

12-5. One mouse is homozygous for *f* (*fT/fT*), and the other mouse is homozygous for *t* (*Ft/Ft*).

12-8. The *AA BB CC* parent produces one kind of gamete – *ABC*. The *aa bb cc* parent also produces one kind of gamete, *abc*. The progeny would all be *AaBbCc*. This will be true whether the genes are linked or not. If the genes are linked, we'd say *ABC/abc*.

12-9. The gametes would be of two kinds of the genes are linked, just like the gametes of the two original parents, *ABC* and *abc*. So if crossed to a parent that is *aa bb cc*, the offspring will be *Aa Bb Cc* and *aa bb cc*. But since the problem specifies that all three loci are linked, the offspring should be symbolized *ABC/ABC* and *abc/abc*.

12-10. Genes are linked when they are on the same chromosome. In this case both vermilion and miniature are on the X chromosome. So not only are they X-linked, they're linked.

12-13. In each case, there will be one parent *cw/cw*. This will only form one kind of gamete: *cw*. The types and proportions of gametes from the heterozygote parent will vary in each instance.

CW/cw X *cw/cw*

	If Independent Assortment	If linked (Heterozygous Male)	If linked (Heterozygous Female)
Parental types	CW/cw (CcWw) 1/4	CW/cw 1/2	CW/cw 0.4
	cw/cw (cc ww) 1/4	cw/cw 1/2	cw/cw 0.4
Recombinant types	cW/cw (cc Ww) 1/4		cW/cw 0.1
	Cw/cw (Cc ww) 1/4		Cw/cw 0.1

An excess of parentals is seen in both cases of linkage, even though there is some recombination in the female.

12-16. The classes that are parental and recombinant would be switched around. The parental gametes would be *Vm* and *vM*, and the cross over gametes would be *VM* and *vm*. The proportions of parentals would still be 97%, and the proportion of cross over gametes would still be 3%.

12-17. Consider the two crosses. Remember that these are X-linked genes.

	Female VM/VM X Male vm/Y		Female vm/vm X Male VM/Y	
Gametes	all (VM)	(vm) (Y)	all (vm)	(VM) (Y)
Offspring: Females	VM/vm		VM/vm	
Males	VM/Y		vm/Y	

Because of the way X-linked inheritance works, only the cross on the right will give the desired result.

CHAPTER 7 LINKAGE MAPS

13-3. The 447 recombinants are 34.7% of the progeny, so the map distance is 34.7.

13-4. In this case the wild type and white miniatures would be the recombinant types. There are 34.7 map units between white and miniature, so 34.7% of the progeny would be these recombinant types.

13-7. s ------- 10 units ------- r ------- 15 units ------- t

13-8.

Distance between the antenna and eye genes	40 map units
Distance between the antenna and bristle genes	10 map units
Distance between eye and bristle genes	45.2 map units

eye ----------- 40 ---------- antenna --- 10 --- bristles

13-13. a) The most common types are MNP and mnp, so those are probably the parental (non-crossover) types. So the heterozygous parent would be MNP/mnp.

b) The heterozygous parent's two homozygous parents are MNP/MNP and mnp/mnp.

c) Divide the problem into three pair-wise problems

MN	409	MP	422	NP	352
Mn	90	Mp	77	Np	127
mN	70	mP	83	nP	153
mn	431	mp	418	np	368

Crossovers
160/1000 =
16% or 16 map units

Crossovers
160/1000 =
16% or 16 map units

Crossovers
280/1000 =
28% or 28 map units

The outside genes are N and P, so the map is:

N--------16--------M--------16--------P

13-14. Arrange the data into three problems of two genes each.

DP	2130	PE	2452	DE	1991
dp	2090	pe	2380	de	1959
dP	550	pE	220	dE	681
Dp	510	Pe	228	De	649

Crossovers
1060/5280 =
20.1 % or 20 map units

Crossovers
448/5280 =
8.5% or 8.5 map units

Crossovers
1330/5280 =
25% or 25 map units

The outside genes are D and E, so the map is

D----------20------------P------8.5------E